Lecture Notes in Chemistry

W0049750

4

Ante Graovac
Ivan Gutman
Nenad Trinajstić

Topological Approach to the Chemistry of Conjugated Molecules

Springer-Verlag
Berlin Heidelberg New York 1977

Authors
Ante Graovac
Ivan Gutman
Nenad Trinajstić
The Rugjer Bošković Institute
P.O.B. 1016
41001 Zagreb, Croatia, Yugoslavia

Library of Congress Cataloging in Publication Data

Graovac, Ante.
    Topological approach to the chemistry of conjugated
molecules.

    (Lecture notes in chemistry)
    Bibliography: p.
    Includes index.
    1. Molecular theory--Mathematical models. 2. Graph
theory. I. Gutman, Ivan, 1947-   joint author.
II. Trinajstić, Nenad, 1936-   joint author.
III. Title.
QD461.G715   541'.22'01514   77-21338

ISBN-13: 978-3-540-08431-0   e-ISBN-13: 978-3-642-93069-0
DOI: 10.1007/ 978-3-642-93069-0

2152/3140-543210

PREFACE

"The second step is to determine constitution, i.e. which atoms
are bonded to which and by what types of bond. The result is expressed by a planar graph (or the corresponding connectivity matrix).... In constitutional formulae, the atoms are represented by
letters and the bonds by lines. They describe the topology of
the molecule."

VLADIMIR PRELOG, Nobel Lecture, December 12th 1975.

In the present notes we describe the topological approach to the chemistry of conjugated molecules using graph-theoretical concepts. Conjugated structures may be conveniently studied using planar and connected graphs because they reflect in the simple way the connectivity of
their pi-centers. Connectivity is important topological property of a
molecule which allows a conceptual qualitative understanding, via a non-
numerical analysis, of many chemical phenomena or at least that part of
phenomenon which depends on topology. This would not be possible solely by means of numerical (molecular orbital) analysis.

Reader´s knowledge of the formal theory of graphs is not essential
for the understanding of the material presented here because these notes are "application-oriented" and the mathematical formalism is reduced to minimum. However, we would like to convince the chemical community at large that the (organic) chemists must develop their knowledge
of mathematics beyond the high-school algebra because many results of
(organic) chemistry may be shaped up nicely by the appropriate mathematical apparatus (group theory, graph theory, theory of polynomials,
functional analysis, etc.) into concise universal laws of chemistry. A
good example to support this statement is provided by the work of Prelog in chemical topology which has led him to introduce the general concept named chirality in order to understand fully the molecular archi-
tecture.

The nomenclature used in these notes has been systematically emplo-
yed and developed by us. We have tried to connect the mathematical terminology with the everyday language of chemists. We shall welcome the
readers´ comments on the material and terminology presented in notes.

Finally, it is our pleasant duty to thank Professor Rudolf Zahrad-
ník,under whose encouragement we started this project, for useful discussions and helpful comments.

Zagreb, Croatia, June 1977.

A. Graovac, I. Gutman, N. Trinajstić

CONTENTS

To Professor BOŽO TEŽAK

1. INTRODUCTION

In these notes we wish to discuss the topological approach to chemistry of conjugated structures using the mathematical apparatus of graph theory (in the text the symbol GT will be sometimes used instead of fully written: graph theory). The basic concepts and definitions of GT will be given in the following chapter. Here, however, we give some general ideas about the use of GT in chemistry.

There is hardly any concept in natural sciences which is closer to the notion of graphs than the structural formulae of chemical compounds. In fact, there is no essential difference between a graph and structural formula.[127,148]

A graph is, simply said, a mathematical structure which may be used to represent the topology of a given molecule. Therefore, chemists can easily grasp the concepts of GT. Moreover, chemists actually know and use a number of graph-theoretical theorems without being aware of this fact in many cases. A classical example is provided by the concept of alternant hydrocarbons introduced by Coulson and Rushbrooke[24] which is for graph-theorists the two-colour problem.[85,149] However, the language of GT is very different from that of chemistry. Therefore, we offer a short glossary in Table 1.1 which should help the reader to follow more easily the text, because we shall freely use and interchange the mathematical and chemical terminology throughout the text.

TABLE 1.1

THE CORRESPONDENCE BETWEEN THE GRAPH-THEORETICAL AND CHEMICAL TERMINOLOGY

Graph-theoretical terminology	Chemical terminology
Molecular (chemical) graph	Structural formula
Vertex	Atom
Weighted edge	Atom of a specified element
Edge	Covalent chemical bond
Weighted edge	Covalent chemical bond between specified elements
Degree of a vertex	Valency of an atom
Tree	Acyclic hydrocarbon
Cycle	Ring
Chain	Linear polyene
Bipartite (bichromatic) graph	Alternant hydrocarbon
Non-bipartite graph	Nonalternant hydrocarbon
Adjacency matrix $\underline{A}$	Topological matrix
Eigenvalue of $\underline{A}$	Molecular orbital energy level
Eigenvector of $\underline{A}$	Topological molecular orbital
Characteristic polynomial	Secular determinant

The advantage of using GT in chemical studies lies in the possibility to apply directly its mathematical apparatus and proof techniques. Besides, a given problem may be considered on a higher level of abstraction which enables a relatively simple insight into the structural features of the molecule. This is a rather important advantage of GT because in many cases we wish to study directly the relations between the particular structural features and a single physicochemical property of a molecule. On the other hand, a purely numerical computerized study sometimes hides the importance of a particular structural feature of a molecule which may account for a molecular property of interest. In addition, the obtained graph-theoretical results have a general validity and may be formulated as theorems and/or rules which can then be applied to any similar group of molecules without any further numerical or conceptual work. Finally, a graph-theoretical language is far more precise and contains a number of terms which have no equivalent in chemistry.

The term molecular topology is appropriately used to describe the non-metric properties of molecules. It should be noted that topology, a branch of mathematics, investigates the nonmetric relationships of geometric (and more abstract) structures. We define molecular topology as the totality of information contained in the molecular graph. Let us emphasize here that the graph-theoretical methods should be expected primarily to be of use as a complementary approach where the topology and the combinatorial nature of problem play an important role, in parallel to the application of the group theory to problems where symmetry is an important feature of the system studied.[103a]

Notes are composed as follows: first, the elements of GT are given and the equivalence between GT and simple molecular orbital theory of conjugated molecules is presented. Then, the pi-electron energy is derived in terms of topological parameters of a molecule. Finally, resonance energies and substitution reactions of conjugated structures are discussed by means of topological theory.

2. GRAPHS IN CHEMISTRY

2.1. BASIC DEFINITIONS AND CONCEPTS OF GRAPH THEORY

2.1.1 Definition of a Graph

There is no unique graph-theoretical terminology. A number of active researchers in this field use their own terms. It is our intention in the present notes to use the terminology of graph theory which we propose for standard use in the chemical literature.

The basic definitions and concepts of graph theory will be given in this section. Since these notes are "application oriented," and thus are designed for the chemical community at large, mathematical rigour is omitted whenever possible. The reader interested in more rigorous definitions, or alternative ones is referred to Refs. 6, 43, 68, and 85.

The Descartes product VxW of sets V and W is defined as

$$VxW = [(v,w) | v \epsilon V, w \epsilon W],\qquad\qquad(2.1)$$

i.e., VxW is the set of ordered pairs, where the first member of the pair is from V and the second one from W. <u>A binary relation R in VxW</u> is any subset of VxW. Especially when V=W, R is called <u>a binary relation (defined) on the set V.</u>

> <u>Example:</u>
> $V = [1,2,3]$
> $VxV = [(1,1), (1,2), (1,3),(2,1), (2,2),(2,3),(3,1),(3,2),(3,3)]$
> $R = [(1,1), (1,2), (2,1), (3,2)]$
> A graph is defined as an ordered pair G
> $$G = (V,R).\qquad\qquad(2.2)$$

Elements of set V are called <u>vertices</u> of a graph and elements of set R are <u>edges</u> of a graph. Often, as well as above, the vertices v_1, v_2,..., v_N will be simply denoted as 1,2,...,N.

In such a way two vertices v_1 and v_2 of the set V either belong to the binary relation

$$(v_1,v_2) \epsilon R\qquad\qquad(2.3)$$

> or they do not

$$(v_1,v_2) \not\epsilon R .\qquad\qquad(2.4)$$

(For an arbitrary ordered pair of vertices either eq. (2.3) or eq. (2.4) is valid.)

> <u>Example:</u>
> $V = [1,2,3,4]$
> $R = [(1,2),(2,4),(3,2),(4,2),(4,3)]$
> $G = (V,R)$

A graph, defined in such an abstract way, can be visualized when the vertices are drawn as small circles and the edge (p,q) is drawn as a <u>oriented (directed) line</u> with the orientation from p to q. Graph G from the above example is represented as G in <u>Fig. 2.1.</u>

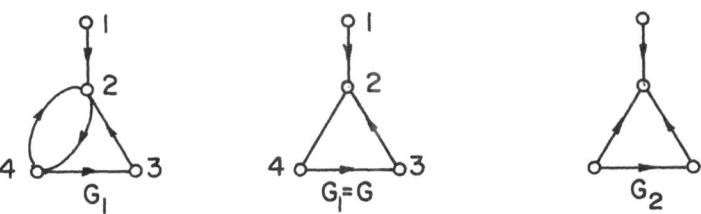

<u>Fig. 2.1.</u> Thē directed graphs.

In graph G a symmetric pair, (2,4) and (4,2), of oriented edges is present, and such a pair can be drawn as an <u>unoriented line (edge)</u> as is done in G_1.

A graph having (only) oriented edges is a <u>directed graph</u> or <u>digraph</u>. An <u>oriented graph</u> is a directed graph having no symmetric pair of directed edges. In <u>Fig. 2.1.</u>, G_1 and G_2 are directed graphs, but only G_2 is oriented.

A graph having all edges unoriented is an <u>unoriented graph.</u>

        <u>Example:</u>

        $V = [1,2,3,4]$

        $R = [(1,2),(2,1),(2,3),(2,4),(3,2),(3,4),(4,2),(4,3)]$

        $G_1 = (V,R)$.

Graph G_1 is represented in <u>Fig. 2.2.</u>

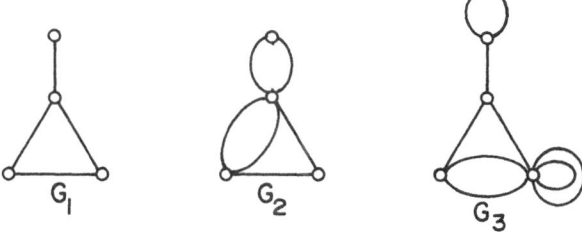

<u>Fig. 2.2.</u> The unoriented graphs.

If more than one edge can join two vertices, we have a <u>multigraph</u>[44] (G_2 in Fig. 2.2.), and if loops (single or multiple) are allowed, we have a <u>pseudograph</u> (G_3 in Fig. 2.2.). A <u>loop</u> is an edge joining a vertex to itself. The orientation of a loop has no significance and therefore is usually omitted.

We will deal with graphs having finite set V of vertices, i.e., with the <u>finite graphs</u> only.

In the following text, if not otherwise stated, a graph will be understood as a finite <u>unoriented</u> graph without loops or multiple edges (G_1 in Fig. 2.2.).

For such a graph G = (V,R) the binary relation R is <u>symmetric</u> and <u>antireflexive</u>, i.e.,

$$(v_1, v_2) \epsilon R \rightarrow (v_2, v_1) \epsilon R \qquad (2.5)$$

and

$$(v_1, v_2) \epsilon R \rightarrow v_1 \neq v_2 . \qquad (2.6)$$

The property (2.5) follows from the requirement for a graph to be undirected, while the property (2.6) reflects the nonexistence of a loop in a graph.

These properties enable us to define an unoriented graph G in an alternative way. We define

$$G = (V,U) \qquad (2.7)$$

where U is a set of unordered pairs of distinct vertices of V, i.e., a set of unoriented edges. The edge joining the vertices p and q is written as $e(p,q)$.

Example:
V = [1,2,3,4]
U = [e_1, e_2, e_3, e_4, e_5]
$e_1 = e(1,2)$, $e_2 = e(2,3)$, $e_3 = e(3,4)$
$e_4 = e(1,4)$, $e_5 = e(1,3)$
$G_1 = (V,U)$.

Graph G_1 is represented in Fig. 2.3.

The total number of vertices and edges of G is denoted by N and M, respectively.

For small N, it is easy to list all possible graphs. In Harary's book[85] the drawings of unoriented graphs with N≤6, as well as those of directed graphs with N≤4, can be found.

The number of unoriented graphs with N≤9(M≤18)[131] and the number of directed graphs with N≤8[122] are also given in Harary's book.[85] The enumeration of graphs is based on Polya's enumeration theorem[125] and later generalizations.

The number of directed graphs far exceeds that of unoriented graphs. As an example, for N=6 there are 156 of that latter type and 1,540,944 of the former.

Vertices p and q are <u>adjacent vertices</u> if an edge e(p,q) is present in G, i.e., if it connects them. Then, we also say that vertex p and edge e(p,q), as well as vertex q and edge e(p,q), are <u>incident</u> with each other. Two edges e and f are <u>incident edges</u> if they have a common vertex.

In the graph G_1 of <u>Fig. 2.3.</u> vertices 1 and 3 are adjacent, but 2 and 4 are not, vertex 1 and edge e_4 are incident, but 1 and e_2 are not, edges e_1 and e_2 are incident, but e_1 and e_3 are not.

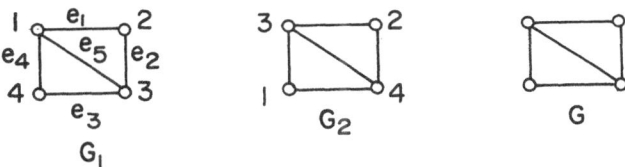

<u>Fig. 2.3.</u> Adjacency and incidence.

In the literature[6,152,164] could be found some others, more general definitions of a graph.

A graph G is <u>labeled</u> if certain numbering of the vertices of G--v_1, v_2,...,v_N-- is introduced. Graphs G_1 and G_2 of <u>Fig. 2.3.</u> are labeled, but G is not.

2.1.2. The Adjacency Matrix of a Graph.
There are several ways of assigning a matrix to a labeled graph.[85]

The most important matrix representation of G is the <u>adjacency matrix</u> <u>A</u> (G) or <u>A</u> (in abbreviated notation). <u>A</u> is the square NxN matrix. The element A_{ij} in the ith row and in the jth column is equal to the number of edges with the orientation from vertex i to vertex j.

As an example, the adjacency matrices <u>A</u>(G_1) and <u>A</u>(G_3) corresponding to the directed graph G_1 of <u>Fig. 2.1.</u> and to the pseudograph G_3 of <u>Fig. 2.2.</u> are:

$$\underline{A}(G_1) = \begin{bmatrix} 0 & 1 & 0 & 0 \\ 0 & 0 & 0 & 1 \\ 0 & 1 & 0 & 0 \\ 0 & 1 & 1 & 0 \end{bmatrix} \qquad \underline{A}(G_3) = \begin{bmatrix} 1 & 1 & 0 & 0 \\ 1 & 0 & 1 & 1 \\ 0 & 1 & 2 & 2 \\ 0 & 1 & 2 & 0 \end{bmatrix} \qquad (2.8)$$

Since we shall consider graphs without multiple edges, we have

$$A_{ij} = \begin{cases} 1 \text{ if } (i,j)\epsilon R \\ \\ 0 \text{ if } (i,j)\notin R \end{cases} \qquad (2.9)$$

and from the nonexistence of a loop in a graph, it follows that

$$A_{ii} = 0. \qquad (2.10)$$

Furthermore, for undirected graphs, the property (2.5) reads as

$$A_{ij} = A_{ji}, \text{ or: } \underline{A} = \underline{A}^+ \qquad (2.11)$$

where $\underline{A}^+$ is the transposed matrix of $\underline{A}$.

In such a way, the adjacency matrix $\underline{A}$ of labeled graph G can be defined alternatively as

$$A_{ij} = \begin{cases} 1 \text{ if } v_i \text{ is adjacent with } v_j \text{ in G} \\ \\ 0 \text{ otherwise} \end{cases} \qquad (2.12.a)$$

or

$$A_{ij} = \begin{cases} 1 \text{ if } e(v_1,v_2)\epsilon U \\ \\ 0 \text{ otherwise.} \end{cases} \qquad (2.12.b)$$

Therefore, the adjacency matrix of a graph is symmetric, its diagonal elements vanish and off-diagonal elements have value 0 or 1, depending on, and reflecting, the connectivity of the vertices in a graph.

For example, the adjacency matrices $\underline{A}(G_1)$ and $\underline{A}(G_2)$ assigned to G_1 and G_2 in Fig. 2.3. are

$$\underline{A}(G_1) = \begin{bmatrix} 0 & 1 & 1 & 1 \\ 1 & 0 & 1 & 0 \\ 1 & 1 & 0 & 1 \\ 1 & 0 & 1 & 0 \end{bmatrix} \qquad \underline{A}(G_2) = \begin{bmatrix} 0 & 0 & 1 & 1 \\ 0 & 0 & 1 & 1 \\ 1 & 1 & 0 & 1 \\ 1 & 1 & 1 & 0 \end{bmatrix} \qquad (2.13)$$

Although G_1 and G_2 obviously represent the same graph (G), their corresponding adjacency matrices differ $[A(G_1)\neq A(G_2)]$ because of different numbering of the vertices.

On the following pages, introduction of new graph-theoretical concepts will be often followed by corresponding matrix formulation. Naturally, those properties and quantities, defined on matrices, which are independent of particular labeling of a graph, will be of the greatest importance.

We can describe the numberings of vertices of a graph by means of permutations. The permutations (1,2,3,4) and P = (3,2,4,1) are ascribed to G_1 and G_2 in Fig. 2.3., respectively. If the permutation P is acting on the columns of $\underline{A}(G_1)$, i.e., if its first column is placed on the position of the third, the third on the place of fourth, the fourth on the place of the first, while the second one does not change its place, and if it is followed by the same permutation procedure on the rows, $\underline{A}(G_2)$ is obtained. This procedure can be put in matrix form by use of permutation matrices.

A permutation matrix $\underline{P}$ is a square NxN matrix defined as follows

$$P_{ij} = \begin{cases} 1 & \text{if mapping of vertices } i \to j \text{ is induced} \\ & \text{by permutation P} \\ \\ 0 & \text{otherwise.} \end{cases} \tag{2.14}$$

It means that in each column and each row of $\underline{P}$ there is only one element equal to 1, and all other elements are equal to 0.

In our case we have

$$\underline{P} = \begin{bmatrix} 0 & 0 & 1 & 0 \\ 0 & 1 & 0 & 0 \\ 0 & 0 & 0 & 1 \\ 1 & 0 & 0 & 0 \end{bmatrix} \tag{2.15}$$

If we multiply $\underline{A}(G_1)$ by $\underline{P}$ from the right side, the columns of $\underline{A}(G_1)$ are permuted

$$\underline{A}(G_1) \cdot \underline{P} = \begin{bmatrix} 1 & 1 & 0 & 1 \\ 0 & 0 & 1 & 1 \\ 1 & 1 & 1 & 0 \\ 0 & 0 & 1 & 1 \end{bmatrix} \tag{2.16}$$

As could be easily shown, multiplication by $\underline{P}^+$ from the left side will lead to the permutation of the rows. Therefore

$$\underline{P}^+ \underline{A}(G_1) \underline{P} = \underline{A}(G_2). \tag{2.17}$$

Because the permutation matrix $\underline{P}$ is a unitary matrix:

$$\underline{P}^+ \underline{P} = \underline{I} \tag{2.18}$$

(where $\underline{I}$ is a unit matrix), the relation between the matrices $\underline{A}(G_1)$ and $\underline{A}(G_2)$ can be rewritten as

$$\underline{P}^{-1} \underline{A}(G_1) \underline{P} = \underline{A}(G_2). \tag{2.19}$$

This result holds generally, i.e., matrices $\underline{A}(G_1)$ and $\underline{A}(G_2)$, corresponding to two numberings G_1 and G_2 of the same graph (G), are similar.

2.1.3.   Isomorphism of the Graphs

We have relied on the reader's intuition in recognizing that the graphs G_1 and G_2 are identical. The procedure for recognizing identical graphs is simple for small graphs like G_1 and G_2 in Fig. 2.3., but it is far more difficult to recognize G_1 and G_2 in Fig. 2.4. as identical graphs.

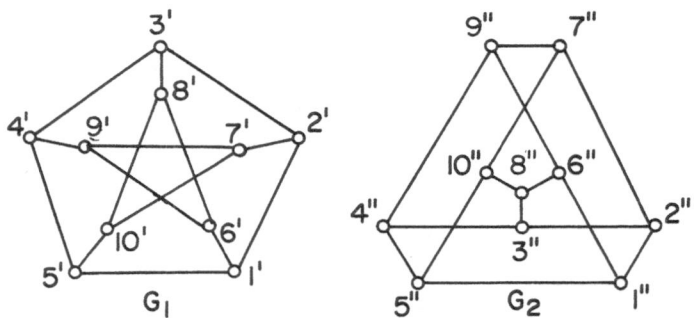

Fig. 2.4.   Isomorphic graphs.

Let us define the isomorphism of graphs. Two graphs $G_1 = (V_1, U_1)$ and $G_2 = (V_2, U_2)$ are isomorphic (we write $G_1 \cong G_2$) if there exists a one-to-one mapping f,

$$fv_1 = v_2 \tag{2.20}$$

such that $(fv_1, fv_2) \epsilon U_2$ if, and only if, $(v_1, v_2) \epsilon U_1$. It means that f needs to preserve adjacency.

Graphs G_1 and G_2 in Fig. 2.4. are isomorphic, because mapping f defined as

$$f \; v_i' = v_i'' \qquad \text{for each i; } (v_i' \epsilon V_1, \; v_i'' \epsilon V_2) \tag{2.21}$$

preserves adjacency.

It follows from the definition that the isomorphic graphs are indeed identical, but differently drawn, graphs.

If graphs are represented by matrices, the isomorphism of graphs can be defined as follows: graphs G_1 and G_2, represented by adjacency matrices $\underline{A}(G_1)$ and $\underline{A}(G_2)$, are isomorphic if and only if there exists a permutation matrix $\underline{P}$ such that   eq. (2.19) is valid.

Because of particular numbering of G_1 and G_2 in Fig. 2.4., it is clear from eq. (2.21) that in our case identity permutation needed is one, i.e., that $\underline{A}(G_1) = \underline{A}(G_2)$.

But generally, the problem of recognizing identical graphs is one of the unsolved problems in graph theory. Construction of all the N! possible mappings (or corres-

ponding permutation matrices) from one graph to another, although obviously impracti-
cal, remains as the only secure check for isomorphism of graphs. The problem of cata-
loguing of chemical compounds stimulated the search for a more practical algorithm.
Sussenguth[147] considered a selection of properties of graphs which form the necessary
conditions for isomorphism of graphs. For a particular class of graphs corresponding
to conjugated systems[128] Randić suggested an algorithm by which an adjacency matrix
of a graph can be brought by permutations to a prescribed standard (canonical) form,
which is defined by a unique numbering of vertices (and which may be of significance
for chemical nomenclature). Identity of graphs is then established by a direct comparison of the corresponding transformed adjacency matrices.

2.1.4.   Further Characterization of a Graph

A graph containing all the vertices of G, but not all the edges of G, is called a
partial graph of G.

    If only one edge e is removed, we write G-e. In Fig. 2.5. G_1 and G-e are partial
graphs of G.

    If both vertices and the edges incident to them are deleted from a graph G, a
subgraph of G is obtained. A special subgraph G-v arises when only one vertex v and
the edges incident with v are removed, while subgraph G-(e) is obtained by deletion
from G of the edge e and its two incident vertices, together with their incident edges.
In Fig. 2.5. G_2, G_3, G-v and G-(e) are subgraphs of G.

    A partial subgraph of G is a partial graph of subgraph of G, i.e., it is obtained
if only edges are deleted in a subgraph. In Fig. 2.5. G_4, G_5 and G_6 are partial subgraphs of G.

    Let us write v_i for a specified vertex and $e = e(v_i, v_j)$ for a specified edge of
a labeled graph G. Then a partial graph G-e is represented by a square NxN matrix
$\underline{A}(G-e)$, which is obtained from adjacency matrix $\underline{A}(G)$ by equaling A_{ij} and A_{ji} to zero.
The same procedure is followed in the case of a general partial graph of G. If we
delete the ith row and ith column of $\underline{A}(G)$, we obtain a square $(N-1) \times (N-1)$ matrix
$\underline{A}(G-v_i)$ which represents a subgraph G-v of G. A subgraph G-(e) is represented by a
square $(N-2) \times (N-2)$ matrix $\underline{A}(G-(e))$, which results when the ith and the jth rows, as
well as the columns are deleted. Other subgraphs, like the partial subgraphs, of G
could be represented in the same manner.

    Let us form a set of all subgraphs $G-v_i$, i= 1,2,...,N for a graph G (with N>3).
If the same set of such subgraphs is obtained for a graph H, the well known Ulam's
conjecture[151] states that G and H are isomorphic graphs. This conjecture is proved
for the trees[106] (the definition follows) and for some other particular class of

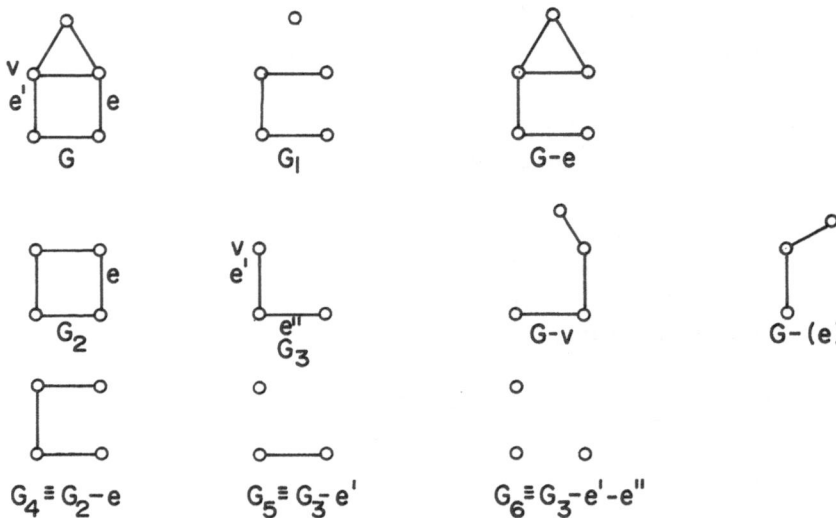

Fig. 2.5. Partial graphs, subgraphs, and partial subgraphs
of a graph.

graphs,[117] but for arbitrary graphs the conjecture remains unsolved.

2.2.   GRAPHS AND TOPOLOGY

2.2.1.   Path, Length and Distance

A path in a graph G is an ordered set of edges $(e_1, e_2, \ldots, e_n)$ with the property that the edge e_j ($1 \leqslant j \leqslant n$) starts from the vertex where the edge e_{j-1} ends. The <u>length</u> of such a path is n. Alternatively, we can declare the order of vertices involved and we write a path as a sequence of vertices $v_0 v_1 v_2 ... v_n$; we call it a v_0-v_n path.

For example, the path (e_1, e_2, e_2, e_5) in G of <u>Fig. 2.6.</u> can be written as 12325.

A <u>closed path</u> is a v-v path, i.e., a path which ends at the same vertex from which it started. Otherwise, a path is <u>open.</u>

Paths 252 and 1234521 in G of <u>Fig. 2.6.</u> are closed, while paths 34 and 45456 are open.

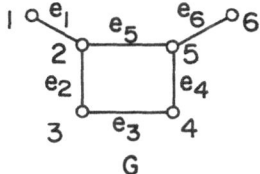

<u>Fig. 2.6.</u> Path, length, and distance on a graph G.

Realize that the paths 1234521 and 34, for example, can be represented by the following products of matrix elements of $\underline{A}$(G): $A_{12}A_{23}A_{34}A_{45}A_{52}A_{21}$ and A_{34}, respectively.

The above formulation can be generalized as follows. A_{pq} can be understood as

$$A_{pq} = \text{the number of paths of length 1 between vertices p and q} \qquad (2.22)$$

and

$$A_{pj}A_{jq} = \begin{cases} 1 \text{ if there is a path of length 2} \\ \quad \text{between vertices p and q passing} \\ \quad \text{through vertex j} \\ \\ 0 \text{ if there is no such a path} \end{cases} \qquad (2.23)$$

Therefore, the expression

$$[A^2]_{pq} = \sum_{j=1}^{N} A_{pj}A_{jq} \qquad (2.24)$$

represents the number of paths of length 2 between vertices p and q. For an arbitrary exponent n, we have

$$A_{pj}A_{jk}\cdots A_{lq} = \begin{cases} 1 & \text{if there is a path of} \\ & \text{length n between vertices} \\ & \text{p and q passing through} \\ & \text{vertices } j,k,\ldots,l \\ \\ 0 & \text{if there is no such a path} \end{cases} \qquad (2.25)$$

and therefore

$$[A^n]_{pq} = \text{the number of paths of length n between} \qquad (2.26)$$
$$\text{vertices p and q.}$$

These considerations will be developed and utilized in the following pages.

The length of the shortest path between two vertices p and q in G is called the distance between these two vertices and is denoted as $d(p,q)$. It is not difficult to see that

$$d(p,q) = 0 \quad \text{if, and only if, } p=q \qquad (2.27)$$
$$d(p,q) = d(q,p) \qquad (2.28)$$
$$d(p,r) + d(r,q) \geqslant d(p,q). \qquad (2.29)$$

Obviously, the distance function d is nonnegative and has only integral values.

The second property (2.28) of d results because we are dealing with an unoriented graph G.

The distance function d defines the metrics on the set of the vertices of G, i.e., the metrics on G.

According to the definition of a graph

$$d(p,q) = 1 \quad \text{if and only if p is adjacent with q in G.} \qquad (2.30)$$

It is easily seen that

$$d(v_i,v_j) = \text{the smallest integer n for which the} $$
$$\text{matrix element } [A^n]_{ij} \text{ is nonzero.} \qquad (2.31)$$

Besides being described by the adjacency matrix $\underline{A}(G)$, the topology of a graph G can be also described by its distance matrix $\underline{D}(G)$, defined as follows:

$$D_{ij} = d(v_i,v_j). \qquad (2.32)$$

It is clear that $\underline{D}$ is a symmetric matrix ($\underline{D}^+ = \underline{D}$) with vanishing diagonal elements. For example, the distance matrix $\underline{D}(G)$ of G of Fig. 2.6. is

$$\underline{D}(G) = \begin{bmatrix} 0 & 1 & 2 & 3 & 2 & 3 \\ 1 & 0 & 1 & 2 & 1 & 2 \\ 2 & 1 & 0 & 1 & 2 & 3 \\ 3 & 2 & 1 & 0 & 1 & 2 \\ 2 & 1 & 2 & 1 & 0 & 1 \\ 3 & 2 & 3 & 2 & 1 & 0 \end{bmatrix} \qquad (2.33)$$

Branching (see later) of a graph is related to the distance matrix,[9,124,155,] since with increasing branching the distances in a graph become smaller.

If every pair of vertices of G is joined by a path, G is a connected graph. If there is no path between two vertices in G (i.e., d=∞), G is a disconnected graph, and these two vertices belong to different components of a graph. All vertices of a graph component have finite distances. The graphs G_1, G_2 and G_3 of Fig. 2.7. have one, two and three components, respectively.

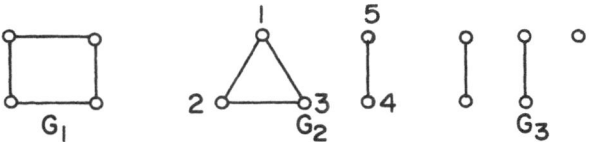

Fig. 2.7. Components of a graph.

For the disconnected graph G it is always possible to label G in such a way that its adjacency matrix $\underline{A}(G)$ has the block-diagonal form. For example, appropriately labeled graph G_2 of Fig. 2.7. has the following adjacency matrix

$$\underline{A}(G_2) = \left[\begin{array}{ccc|cc} 0 & 1 & 1 & 0 & 0 \\ 1 & 0 & 1 & 0 & 0 \\ 1 & 1 & 0 & 0 & 0 \\ \hline 0 & 0 & 0 & 0 & 1 \\ 0 & 0 & 0 & 1 & 0 \end{array} \right] \qquad (2.34)$$

It is evident that the number of components of G is equal to the number of block-matrices of $\underline{A}(G)$, while the number of rows and columns in particular block-matrix equals the number of vertices in the corresponding component of G.

2.2.2. Neighbours. The Invariants of a Graph

Once the distance function is defined, it is easy to introduce the notion of neighbours. All vertices among which the distance is 1 (i.e., the adjacent vertices) are the first neighbours (or simply neighbours). The second, the third and the higher neighbours are defined in an analogous way.

The number of the first neighbours of a vertex p is called the <u>degree</u> of a vertex p, and we write $d_1(p)$ or simply $d(p)$. The number of the second, the third,..., neighbours of vertex p will be denoted by $d_2(p)$, $d_3(p)$ ... .

For G_2 of <u>Fig. 2.7.</u> we have: $d(1) = d(2) = d(3) = 2$, $d(4) = d(5) = 1$.

Obviously, $d(p)$ equals the number of edges incident with p. Therefore

$$\sum_{i=1}^{N} d(v_i) = 2M \tag{2.35}$$

(where M = the number of edges of G) because in the summation each edge is counted twice. From eq. (2.35) follows that for any graph the number of vertices of odd degree is even.

Let us denote the number of vertices having $d(p) = 1,2,3,...$ by P, S, T, ... (evidently: P+S+T+...=N). Then the following identity is valid

$$P + 2S + 3T + ... = 2M . \tag{2.36}$$

We can express $d(v_i)$ as the sum of elements in the ith row of the adjacency matrix $\underline{A}(G)$

$$d(v_i) = \sum_{j=1}^{N} A_{ij} \tag{2.37}$$

and also as

$$d(v_i) = [A^2]_{ii} = \sum_{j=1}^{N} A_{ij}A_{ji} \tag{2.38}$$

because: $A_{ij} = A_{ij}^2 = 0$ or 1, and: $A_{ij} = A_{ji}.$

The <u>trace</u> Tr $\underline{C}$ of a square NxN matrix $\underline{C}$ is defined as the sum of its diagonal elements

$$\text{Tr } \underline{C} = \sum_{i=1}^{N} C_{ii} . \tag{2.39}$$

Different labeling of the same graph results in different adjacency matrices, which are related through eq. (2.19). When the <u>cyclic property of the trace</u>

$$\text{Tr}(\underline{ABC}) = \text{TR}(\underline{BCA}) = \text{Tr}(\underline{CAB}) \tag{2.40}$$

is used, it follows that the trace is independent of labeling and we write Tr $\underline{A}$. We say Tr $\underline{A}$ is <u>invariant</u> of a graph, i.e., this quantity is characteristic of a graph, as well as N, M, the set of degrees of vertices, and so on.

We have

$$\text{Tr } \underline{A} = 0 \tag{2.41}$$

what reflects the nonexistence of loops in a graph.

The trace of powers of $\underline{A}$ will be invariant too, as can be seen for $\underline{A}^2$:

$$\text{Tr } \underline{A}^2 = \sum_{i=1}^{N} d(v_i) = 2M. \tag{2.42}$$

Although the complete set of invariants of a graph is not known, it is obvious that these quantities would play a role of special importance in graph-theoretical considerations throughout the present notes.

For a directed graph the notion of <u>outdegree</u> and of <u>indegree</u> of a vertex p is introduced, the first being the number of directed edges coming out from p (we write od(p)) and the second being the number of directed edges entering p (we write id(p)).

2.2.3. <u>Ring and Oriented Ring. Regular and Complete Graphs. The Ring and the Edge Components of a Graph</u>

If all vertices in a connected graph G are of degree 2, a graph is called a <u>ring</u> (or a <u>cycle</u>). It is evident that the corresponding adjacency matrix $\underline{A}$(G) contains in each row (as in each column) two elements equal to 1, and therefore the number of edges (M) equals the number of vertices (N). The graphs G_1, G_2 and G_3 of <u>Fig. 2.8.</u> are the rings with 3,4 and 5 vertices, respectively.

The rings are special case (d = 2) of <u>regular graphs of degree d</u>, which are defined as graphs having all vertices of the same degree: $d(v_1) = d(v_2) = \ldots = d(v_N)$ = d. The graphs G_4 and G_3 of <u>Fig. 2.8.</u> are regular graphs of degree 3.

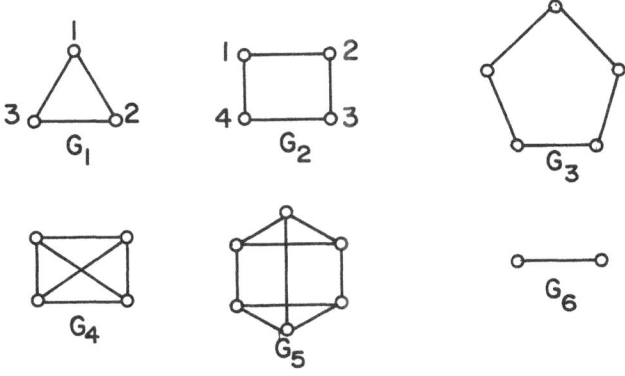

<u>Fig. 2.8.</u> Rings. Regular graphs. Edge.

For a regular graph G of degree d eq. (2.42) reads

$$M = Nd/2 \qquad\qquad (2.43)$$

because the corresponding adjacency matrix $\underline{A}(G)$ contains d elements equal to 1 in each row (as well as in each column). From the above equation it follows that a regular graph exists only if N or (and) d are even.

If d = 1, there exists only one connected regular graph. It is G_6 in <u>Fig. 2.8.</u> and we call it the edge.

A regular graph G with N vertices and d = N-1 is called a <u>complete graph</u> and is denoted by K_N. In such a graph the edges are connecting each possible pair of vertices and therefore

$$M = \binom{N}{2} = \frac{N(N-1)}{2} \ . \qquad\qquad (2.44)$$

The same result is obtained starting from eq. (2.42), when we realize that in each row (altogether N of them) of $\underline{A}(K_N)$ all elements, except those on the diagonal, are equal to 1.

G_6 (the edge), G_1 and G_4 of <u>Fig. 2.8.</u> are the complete graphs K_2, K_3 and K_4, respectively.

If in an oriented graph each vertex p has indegree id(p) = 1 and outdegree od(p) = 1, we call it an <u>oriented ring</u>. G_1, G_2, G_3, G_4 and G_5 of <u>Fig. 2.9.</u> are oriented rings with N = 3,3,4,4 and 2, respectively. Although, G_5 is the same as the edge (G_6) of <u>Fig. 2.8.</u>, it is included here because of unique treatment and because of reasons which will become clearer later.

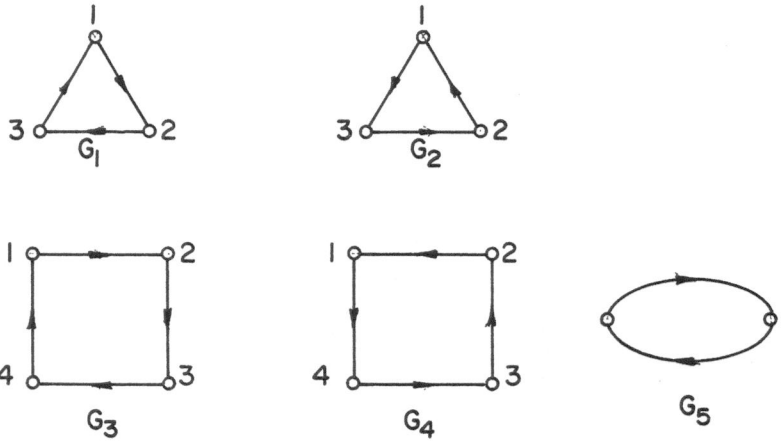

<u>Fig. 2.9.</u> The oriented rings.

A ring with N vertices will be also called an <u>N-membered ring.</u> The number of 3-,4-,...,j-,... membered rings, present in a graph G, will be denoted by $n_3(G)$, $n_4(G)$, ..., $n_j(G)$,..., respectively, or simply by n_3, n_4,...,n_j,... . In <u>Fig. 2.10.</u> we have graphs with $n_3 = 2$, $n_4 = 1$, $n_5 = n_6 = \ldots = 0$.

Previously, eqs. (2.41) and (2.42) for Tr $\underline{A}$ and Tr $\underline{A}^2$ were given. In the same manner the trace of higher powers of $\underline{A}^n$ can be expressed by counting the number of "structural details" of G. It is easy to prove that

$$\text{Tr } \underline{A}^3 = 6n \qquad\qquad (2.45)$$

and

$$\text{Tr } \underline{A}^4 = \sum_{i=1}^{N} d^2(i) \; + \sum_{i=1}^{N} d_2(i) \; + 6n_3 \; + 12n_4, \qquad (2.46)$$

but for n>4 the expressions become quite unwieldy.

Similar considerations will be utilized in the following sections.

2.2.4. <u>Sachs Graphs with N Vertices</u>

For an oriented ring G with N vertices, let us form the following product of N elements of $\underline{A}(G)$:

$$A_{1j_1} A_{2j_2} \ldots A_{ij_i} \ldots A_{Nj_N} \qquad\qquad (2.47)$$

where $(j_1 j_2 \ldots j_i \ldots j_N)$ is some permutation of the starting permutation $(12 \ldots i \ldots N)$, i.e., in the product one and only one element from each row and column of $\underline{A}(G)$ is taken into account. There are N! products of the above type. If for at least one factor A_{ij_i} there is no oriented edge $e(i,j_i)$ in G, the product vanishes. Therefore only one product will have nonzero value (which is equal to 1). It is the product which corresponds to the "circulation" around the ring, starting from some vertex, following the direction of edges and ending at the starting vertex. For oriented rings G_1, G_2, G_3, G_4 and G_5 in <u>Fig. 2.9.</u>, these nonzero products are $A_{12}A_{23}A_{31}$, $A_{13}A_{32}A_{21}(=A_{12}A_{23}A_{31})$, $A_{12}A_{23}A_{34}A_{41}$,$A_{14}A_{43}A_{32}A_{21}(=A_{12}A_{23}A_{34}A_{41})$ and $A_{12}A_{21}$, respectively.

Similar considerations can be also done for (unoriented) rings. But because in such a ring the circulation around it can be undertaken in two opposite directions (clockwise and counterclockwise), two products of type (2.47) will have nonzero value. For G_1 of <u>Fig. 2.8.</u> these products are $A_{12}A_{23}A_{31}$ and $A_{13}A_{32}A_{21}$, which correspond to the oriented rings G_1 and G_2 of <u>Fig. 2.9.</u>

For the edge (N=2), only one nonzero product will arise $(A_{12}A_{21})$.

The above considerations can be further generalized to an arbitrary graph G. For instance, there are four nonzero products of type (2.47) for G in <u>Fig. 2.10.</u>: $A_{12}A_{21}$

$A_{34}A_{45}A_{56}A_{63}$, $A_{12}A_{21}A_{36}A_{65}A_{54}A_{43}$, $A_{12}A_{21}A_{36}A_{63}A_{54}A_{45}$, and $A_{12}A_{21}A_{34}A_{43}A_{56}A_{65}$, the first two correspond to G_1 (or alternatively to G_1' and G_1''), while the second and the third correspond to G_2 and G_3.

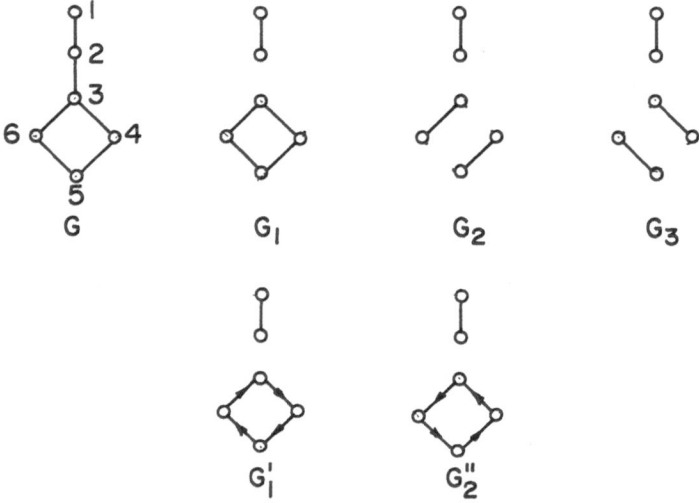

Fig. 2.10.   The ring and edge components. The oriented ring
components.   The Sachs graphs.

In such a way, the search for nonvanishing products of N elements, defined on a graph G by expression (2.47), led us to the particular class of partial graphs of G. Each of them contains disconnected <u>ring</u> and <u>edge components</u> of G only. It is called a <u>Sachs graph with N vertices</u>[52,138] and is denoted by s. The set of all Sachs graphs with N vertices of G is denoted by $S_N(G)$, or simply by S_N. As example, G in <u>Fig. 2.10.</u> has $S_6 = [G_1, G_2, G_3]$.

The number of ring components of s is denoted by $r(s)$ and the number of ring and edge components (altogether) of s is denoted by $c(s)$. Sachs graph G_1 in <u>Fig. 2.10.</u> has $r(s) = 1$ and $c(s) = 2$, while G_2 and G_3 have $r(s) = 0$ and $c(s) = 3$.

Sachs graph s corresponds to some nonvanishing products of type (2.47). Each edge component occurs once, while each ring component occurs twice as a factor in such products. Therefore,

<blockquote>
Sachs graph s ($s\epsilon S_N$) with $r(s)$ ring components describes
$2^{r(s)}$ nonvanishing products of type (2.47)   (2.48)
</blockquote>

and

each of such products has the value equal to 1.   (2.49)

Similar considerations will play an essential role in the following sections concerning Sachs theorem and related generalizations.

A notion of <u>oriented Sachs graph</u> $(\vec{s})$ will be introduced now. Such a graph contains disconnected <u>oriented ring components</u> of (unoriented) graph G only. From the definition of oriented ring, it follows that

oriented Sachs graph $\vec{s}$ $(\vec{s} \epsilon \vec{S}_N)$ corresponds to one and

only one nonvanishing product of type (2.47)   (2.50)

and

such product equals 1.   (2.51)

For (unoriented) graph G in <u>Fig. 2.10.</u> the set of all oriented Sachs graphs $(\vec{S}_N)$ is $\vec{S}_6 = [G_1', G_1'', G_2, G_3]$.

When more rings in G are present, the number of oriented Sachs graphs far exceeds that of unoriented Sachs graphs, and therefore, between these two equivalent representations, we shall prefer the use of unoriented Sachs graphs.

Until now we have introduced the most important notions about graphs. Further useful notions like planarity of graphs, the colouring of graphs, etc., will be introduced in the forthcoming sections as necessary.

2.3   GRAPHS REPRESENTING CONJUGATED MOLECULES

2.3.1   Planar Graphs. Colouring of Graphs

A graph can be drawn in many different ways. As example, G_1 and G_2 ($\cong G_1$) in **Fig. 2.11.** are just two different representations (in the plane) of the same graph. In representing G_2 no two edges intersect. We say a graph is <u>planar</u> if it can be drawn in the plane in such a way that no two edges intersect. G_1 (and G_2) in **Fig. 2.11.** is a planar graph, while G_3 (the complete graph K_5) and G_4 (the complete bigraph $K_{3,3}$, see later) and G_5 are not. G_3 and G_5 are the smallest nonplanar graphs.

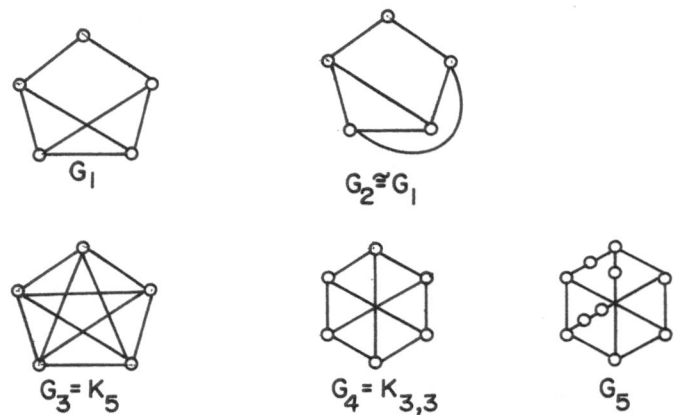

Fig.2.11. Planar and nonplanar graphs. Homeomorphs of a graph.

As the result of the <u>Euler polyhedron formula</u> (indeed planar graphs were introduced by Euler in his studies on polyhedra), it follows[85] that in a planar graph at least one vertex of degree less that 6 exists.

If we place new vertices on the edges of a graph G, a <u>homeomorph</u> of G is obtained. G_4 and G_5 of <u>Fig. 2.11.</u> are <u>homeomorphic.</u>

Kuratowski proved[110] that a graph is planar if and only if it has neither K_5, $K_{3,3}$, nor their homeomorphs as its partial subgraphs.

The <u>four colour problem,</u> stated in the 1850s, stimulated fruitful interest in problems of the colouring of graphs. <u>Colouring of a graph</u> is carried out in such a way that its adjacent vertices are coloured with different colours; if n colours are used we have n-colouring of a graph. It is obvious that <u>n-colouring</u> of a graph G (with N vertices) can always be done. If we decrease the number of colours used, we will finally meet the number k, called the <u>chromatic number</u> k = chi(g), such that k-colouring of a graph is possible but not (k-1)-colouring. Then G is <u>k-chromatic.</u>

As example, in Fig. 2.12. there is shown one 6-colouring (a) and 3-colouring (b) of the same 3-chromatic (trichromatic) graph $G_1(N = 6)$. The number in parentheses indicates the number of colours used.

The important class of graphs is 2-chromatic (bichromatic) graphs as, for example, G_2 and G_3 in Fig. 2.12. The colouring process for bichromatic graphs will be indicated by stars (*) and circles (o) as shown for G_2. Vertices of different colours in bichromatic graphs will be therefore called starred and unstarred. Conventionally

$$p \geq q \qquad (2.51)$$

where p and q denote the number of starred and unstarred vertices, respectively. Obviously: p + q = N.

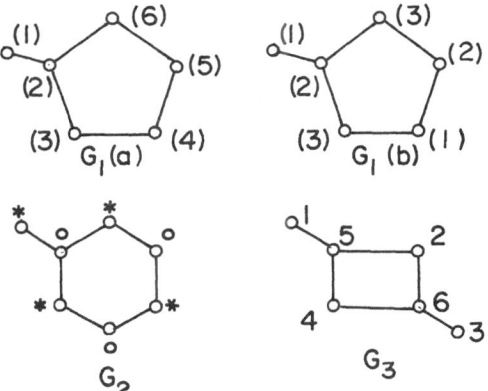

Fig. 2.12. Colouring of graphs. Trichromatic and bichromatic graphs.

The following theorem[108] is valid: a graph is bichromatic if and only if there is no odd-membered ring component of a graph. Therefore, there is no difficulty in deciding by inspection whether one graph is bichromatic or not, but generally the determination of chi(G) is not possible without an effective colouring algorithm.[164]

If in a bichromatic graph G there exists every edge connecting starred (p of them) and unstarred (q of them) vertices, G is a complete bichromatic graph and we write $G = K_{p,q}$. As an example, G_4 of Fig. 2.11. is the $K_{3,3}$ graph.

If we label (the vertices of) a bichromatic graph in such as way that $1, 2, \ldots, p$ are starred and $p+1, p+2, \ldots, p+q=N$ are unstarred vertices, it is obvious that

$$A_{ij} = 0 \quad \text{for } 1 \leq i, j \leq p \quad \text{and for } p+1 \leq i, j \leq p+q \qquad (2.52)$$

because two starred, as well as unstarred, vertices are not adjacent. Therefore, for the above labeling the adjacency matrix $\underline{A}$ has the form[84]

$$\underline{A} = \begin{bmatrix} \underline{0} & \underline{B} \\ \\ \underline{B}^+ & \underline{0} \end{bmatrix} \qquad (2.53)$$

where $\underline{B}$ is (pxq) submatrix and $\underline{B}^+$ is (qxp) transposed matrix of $\underline{B}$.

As an example, for the labeled bichromatic graph G_3 of Fig. 2.12. we have

$$A(G_3) = \begin{bmatrix} 0 & 0 & 0 & 0 & 1 & 0 \\ 0 & 0 & 0 & 0 & 1 & 1 \\ 0 & 0 & 0 & 0 & 0 & 1 \\ 0 & 0 & 0 & 0 & 1 & 1 \\ \hline 1 & 1 & 0 & 1 & 0 & 0 \\ 0 & 1 & 1 & 1 & 0 & 0 \end{bmatrix} \qquad (2.54)$$

Often the terms bipartite and tripartite graphs are used for bichromatic and trichromatic graphs.

2.3.2. Hückel Graphs

Let us remind the reader that a graph was defined by a set of vertices and by a binary relation (a set of edges) defined on the set of vertices. One property of molecules is very close to a binary relation, namely, two atoms in molecule are either bonded or not bonded. Therefore, molecules can be represented by graph when only the possible existence of a chemical bond between the atoms in a molecule is considered, while all other molecular properties are neglected. When atoms are represented by vertices and bonds by edges a molecular graph is obtained. The analogy between structural formulae and graphs is obvious. It is illustrated on the example of acidic acid (a) in Fig. 2.13.

But the graphs which describe molecules can be introduced in a different way, following the quantum-chemical scheme for the construction of the molecular orbitals[126] The procedure is illustrated on the example of methane (b) in Fig. 2.13. First a basis graph (G_2 in Fig.2.13.) is constructed,i.e.,a graph in which vertices correspond

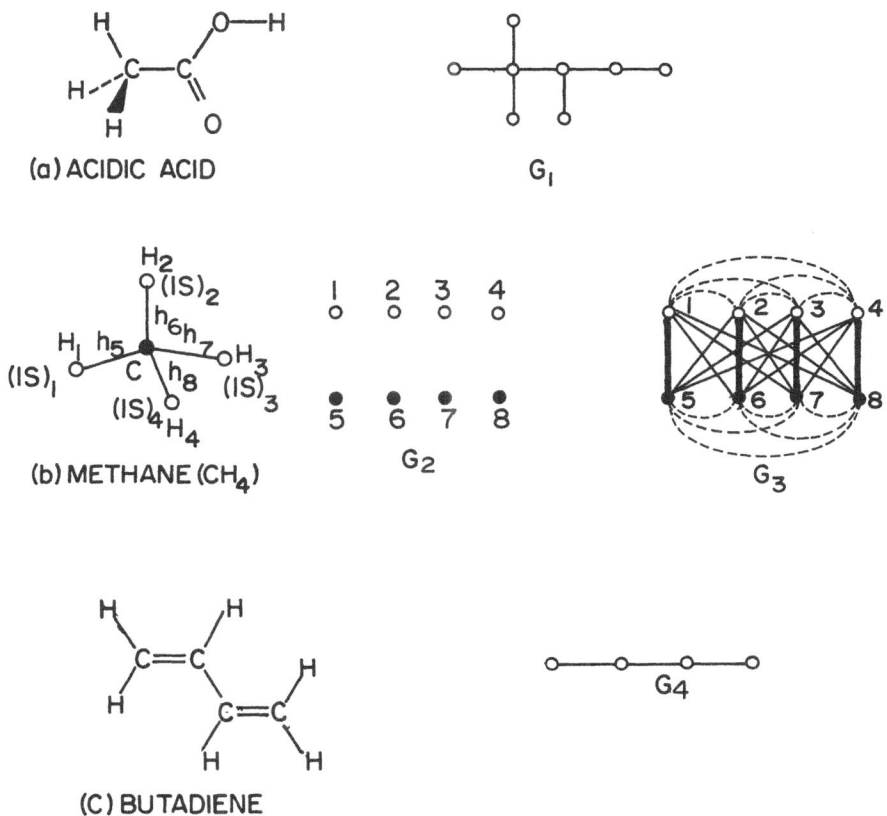

(a) ACIDIC ACID

G_1

(b) METHANE (CH$_4$)

G_2

G_3

(C) BUTADIENE

G_4

Fig. 2.13.   The molecular graph.   The atomic orbital graph.
The quantum chemical graph.   The Hückel graph.

to basis functions used. In our example, 1s atomic orbitals of H-atoms are represented
by the vertices of one particular type (0), and sp^3 hybrid orbitals of C-atom by the
vertices of another type (●). Then the Hamiltonian matrix of a problem needs to be
formed. The various interactions between the orbitals used (usually expressed by
means of the molecular integrals) can be described by (the vertices and) the edges
of different weights. The corresponding <u>weighted graph</u> which is G$_3$ of <u>Fig. 2.13.</u> is
called the <u>quantum chemical graph</u>. In our example, because of the symmetry of the
methane and the atomic orbitals used, only four different interactions are present.
Their representatives are $(1s)_1$-h_5, $(1s)_1$-h_6, $(1s)_1$-$(1s)_2$ and h_5-h_6 interactions
which are denoted by the edges of four different types: ○━●, ○──●, ○···○, and ●···●,
respectively. Later, we shall prefer to use the numbers, rather than the vertices

and the edges of different types, to represent the different weights of the vertices
and the edges.

Now, the important class of graphs will be introduced. When conjugated hydrocarbons are considered, the related graphs correspond to the carbon-carbon sigma-
bond skeleton, while the H-atoms, pi-bonds and C-H sigma-bonds are neglected. For
example, butadiene is represented by the graph G_4 as shown in Fig. 2.13. In the
present notes all graphs are assumed to correspond to conjugated hydrocarbons in this
way.

The graphs considered are planar and the degree of their vertices does not ex-
ceed 3. They are called Hückel graphs[53],[54] to distinguish them from the other types
of molecular graphs. They represent the pi-electron network of a given conjugated
molecule. We see that they could be understood as the simple particular case of
quantum-chemical graphs with only pi-interactions considered.

2.3.3.   Trees. Benzenoid Graphs

The acyclic structures of organic chemistry are described by trees. A tree is a connected acyclic graph. If it has N vertices, it contains N-1 edges. Obviously, by
cutting any of its edges, a tree can be separated into two parts. Some trees with 7
vertices are shown in Fig. 2.14.; G_3 is not a Hückel tree. In Harary's book[85] the
drawings of all the trees with N≤10 vertices, as well as the number of them for N≤26,
could be found.

A tree possesses necessarily vertices of degree one. Such vertices are called
terminals. The tree with the minimal number of terminals (two of them) is the chain
(as it is G_4 in Fig. 2.13.), while the one with the maximal number (n-1) of terminals
is the star (as it is graph G_3 in Fig. 2.14.).

It is obvious that the trees belong to the bichromatic graphs.

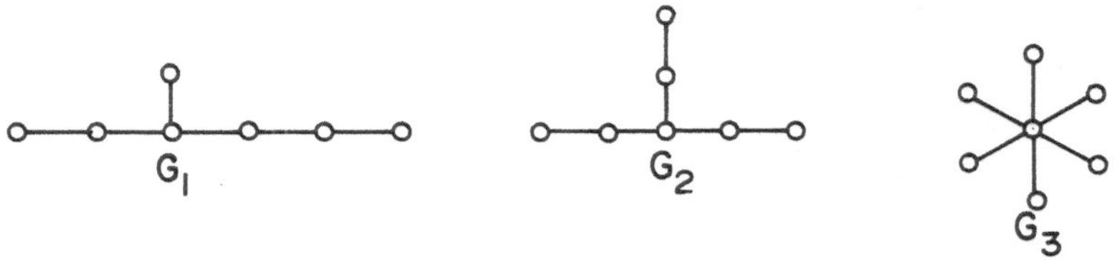

Fig. 2.14.   Some trees with N = 7.

Because the trees are graphs of relatively simple structure, very often various graph-theoretical methods and concepts are first tested on them.

For instance, the branching of a graph, the factor determining a number of molecular properties[66,129] could be best studied for the trees.[9a] A measure of branching needs to be discriminate in a unique way between, for example, the graphs G_1, G_2 and G_3 in Fig. 2.14. It must necessarily fulfill a condition that it has the minimal value for a chain and the maximal value for a star, or vice versa.

As the second example, the enumeration of graphs, a problem related to isomer enumeration, was also first studied for the trees.[15] Isomer enumeration methods are based on Polya's theorem.[125,134]

We will now define another special class of Hückel graphs which are of great importance for chemistry since they represent benzenoid hydrocarbons. These are benzenoid graphs which are obtained by any combination of regular hexagons (in a plane) such that two hexagons have either exactly one common edge or are disjoint. The examples (triphenylene and perylene graphs) are given in Fig. 2.15.

If we join the centers of adjacent pairs of hexagons in a benzenoid graph, as shown in Fig. 2.15., its dual or characteristic graph is obtained. If the characteristic graph obtained is a tree, the molecule is cata-condensed (G_1), while if it is not a tree, the molecule is peri-condensed (G_2 of Fig. 2.15.).

This definition can be introduced in many other alternative ways. A molecule is called cata-condensed if the number of internal vertices equals zero; or if no vertex belongs to three rings; or if no three rings are mutually adjacent; or if all vertices belong to the perimeter; or if all cycles contained in a molecular graph are of the length $4m + 2$.

A (benzenoid) graph divides the plane into one infinite and some number of finite regions (rings). All vertices and edges which lie on the boundary of the infinite region form the perimeter of a graph. The internal vertices are those not belonging to the perimeter. For G_2 in Fig. 2.15., they are labeled by ●.

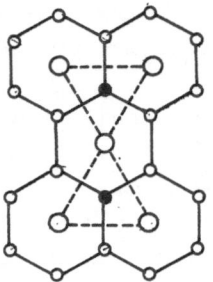

Fig. 2.15. Benzenoid graphs.

The problem of enumeration of benzenoid graphs is studied[86] and their topological properties are investigated[57]

2.4 GRAPH SPECTRUM. SACHS THEOREM

2.4.1. Graph Spectrum

The adjacency matrix $\underline{A}$ of a graph G could be submitted to various transformations. The transformation leading to the diagonal matrix form

$$\underline{X} = \begin{bmatrix} x_1 & & & & \\ & x_2 & & \bigcirc & \\ & & \ddots & & \\ \bigcirc & & & & x_N \end{bmatrix} \tag{2.55}$$

is of the utmost importance. N is the number of vertices of G.

Diagonal elements x_1, x_2, ..., x_N of $\underline{X}$ are called the eigenvalues of the matrix $\underline{A}$. Conventionally,

$$x_1 \geqslant x_2 \geqslant \cdots \geqslant x_N. \tag{2.56}$$

If some eigenvalue appears k times, we say it is k-fold degenerate.

The set of all eigenvalues $[x_1, x_2, \ldots, x_N]$ is called the spectrum of a graph G (graph spectrum).

Because the adjacency matrix $\underline{A}$ for an (unoriented) graph is symmetric, the above-mentioned diagonalization can be always carried out and furthermore x_j (j = 1,2,...,N) are real numbers[11] The transformation to diagonal form is realized by the eigenvector matrix $\underline{C}$ of $\underline{A}$

$$\underline{C}\,\underline{A} = \underline{X}\,\underline{C} \tag{2.57}$$

where $\underline{C}$ is NxN matrix of the form

$$\underline{C} = \begin{bmatrix} \underline{C_1} \\ \underline{C_2} \\ \cdot \\ \cdot \\ \cdot \\ \underline{C_N} \end{bmatrix}. \tag{2.58}$$

The row 1xN matrices $\underline{C_j}$

$$\underline{C_j} = (C_{j1}, C_{j2}, \ldots, C_{jN}) \tag{2.59}$$

obey

$$\underline{C_j}\,\underline{A} = x_j\,\underline{C_j}, \quad j = 1,2,\ldots,N, \tag{2.60}$$

and we say the $\underline{C}_j$ is eigenvector of $\underline{A}$ with the eigenvalue x_j.

As shown earlier (see eq. (2.19)), the adjacency matrices corresponding to two different numberings of the same graph are similar and therefore, according to the well-known result from matrix algebra,[111] they have the same eigenvectors and eigenvalues. Hence, the spectrum of a graph is independent of the numbering of its vertices, i.e., it is a graph invariant. It plays the central rôle in our presentation. The formal spectral theory of graphs was introduced[18] in 1957, but in quantum chemistry the subject has been known (at least implicitly) for 45 years,[100] thus showing once more how true are the words of mathematician Sylvester:[148] "There is an untold treasure of hoarded algebraic wealth potentially contained in the results achieved by the patient and long-continued labour of our unconscious and unsuspected chemical fellow workers."

Several review articles on the spectral methods in the graph theory may be found in the literature.[25a,73]

Obviously, the isomorphic graphs are isospectral or cospectral, i.e., they have the same spectra. But the isospectral graphs are not necessarily isomorphic. The simplest examples of such graphs interesting for chemists are G_1 and G_2 of Fig. 2.16. The subject has been investigated by several authors.[4,68,90a,91a,130,166] It is shown[150] that not only the spectrum of the adjacency matrix $\underline{A}$, but also the spectrum of more general matrix functions of a graph do not determine a graph up to isomorphism.

Although a graph is not uniquely determined by its spectrum, nevertheless the graph spectrum yields several pieces of information about the structure of a graph; more valuable information is obtained[25a] from study of a narrower class of graphs.

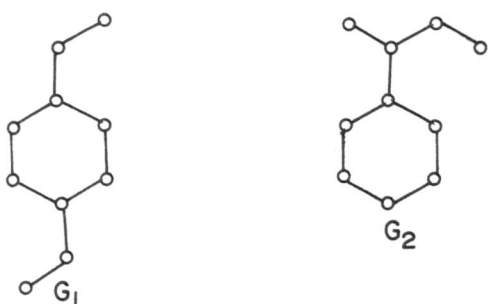

Fig. 2.16.   The pair of isospectral nonisomorphic graphs.

However, we will be more interested to see how the structure (topology) of a graph influences its spectrum. Let us examine one of the simplest and the most obvious relationships. Because the matrices $\underline{A}$ and $\underline{X}$ are similar and therefore their traces are the same for any power of $\underline{A}$ and $\underline{X}$, eqs. (2.41), (2.42), (2.45), and (2.46) can be rewritten as follows:

$$\text{Tr } \underline{A} = \sum_{j=1}^{N} x_j = 0 \tag{2.61}$$

$$\text{Tr } \underline{A}^2 = \sum_{j=1}^{N} x_j^2 = 2M \tag{2.62}$$

$$\text{Tr } \underline{A}^3 = \sum_{j=1}^{N} x_j^3 = 6n_3 \tag{2.63}$$

$$\text{Tr } \underline{A}^4 = \sum_{j=1}^{N} x_j^4 = \sum_{j=1}^{N} d^2(j) + \sum_{j=1}^{N} d_2(j) + 6n_3 + 12n_4 \tag{2.64}$$

thus indicating that the structure of a graph influences the graph spectrum.

The procedure for evaluation of the (graph) spectrum and the corresponding eigenvectors is well known in matrix algebra.[11,121] Eq. (2.60) can be rewritten in the form

$$\underline{C}_j \ (x_j \ \underline{I} - \underline{A}) = 0 \tag{2.65}$$

(where $\underline{I}$ stands for unit matrix), i.e., as a system of linear equations, so-called secular equations

$$\sum_{k=1}^{N} C_{jk} \ (x_j \ \delta_{kl} - A_{kl}) = 0, \quad 1 = 1, 2, \ldots, N. \tag{2.66}$$

In order to have a (nontrivial) solution of eqs. (2.66) it is necessary that the corresponding secular determinant vanish

$$\det \ (x \ \underline{I} - \underline{A} \ ) = 0 \tag{2.67}$$

The polynomial

$$P(G;x) = \det \ (x \ \underline{I} - \underline{A} \ ) \tag{2.68}$$

is called the characteristic polynomial of a graph G and it is also a graph invariant. $P(G;x)$ is the polynomial of the degree N

$$P(G;x) = \sum_{j=0}^{N} a_j \ x^{N-j} \tag{2.69}$$

where a_j's are the coefficients of the characteristic polynomial.

The set of roots of the characteristic polynomial of a graph forms the graph spectrum. For a particular root (eigenvalue) x_j, eqs. (2.66) will give the corresponding eigenvector $\underline{C}_j$. From the definition (2.68) it follows that

$$a_0 = 1 \qquad (2.70)$$

and we can write $P(G;x)$ as

$$P(G;x) = (x-x_1)(x-x_2) \ldots (x-x_N). \qquad (2.71)$$

It is easily shown[52,143] that

$$a_1 = \sum_{j=1}^{N} x_j = 0 \qquad (2.72)$$

$$a_2 = -\tfrac{1}{2} \sum_{j=1}^{N} x_j{}^2. \qquad (2.73)$$

Comparison with eqs. (2.61) and (2.62) indicates that the coefficients of $P(G;x)$ will also be dependent on the structure (topology) of a graph G. Such considerations will be fully developed in paragraph 2.4.3.

2.4.2.   Graph Spectral Properties of Particular Classes of Graphs

For certain classes of graphs the graph spectrum (and the eigenvector problem) can be solved in a closed analytical form, most frequently using the symmetry properties of a graph. Often, in analysis of a particular class of graphs (G), which are obtained when two (smaller) graphs (G_1 and G_2) are connected with only one edge e (v_1, v_2), it is useful to apply the formula of Heilbronner[87]

$$P(G;x) = P(G_1;x) P(G_2;x) - P(G_1-v_1;x) P(G_2-v_2;x). \qquad (2.74)$$

Using these and similar methods the graph spectral-structural analysis was carried out analytically for regular and complete graphs[25a] especially for rings[100] for chains[100] stars and other trees[60] for radialenes and related molecular graphs[76] as well as for some other classes of graphs[25a]

The numerical data for trees and other graphs were calculated by Nosal[121a] while those for the most important Hückel graphs can be found, for example, in Coulson and Streitwiesser's tables[25]

The interval in which the eigenvalues of a graph lie is limited. According to Frobenius theorem[48]

$$-d_{max} \leqslant x_j \leqslant d_{max}, \quad j = 1,2,\ldots,N, \qquad (2.75)$$

where d_{max} is the maximal vertex degree in a graph. Therefore, for the Hückel graphs[21,115] the whole spectrum lies in the interval from -3 to +3.

The largest eigenvalue (x_1) of a graph depends on a branching of its skeleton[66,115]

The eigenvalue with the smallest absolute value (x_{min}) is related to the HOMO-LUMO (highest occupied-lowest unoccupied molecular orbital) separation of related conjugated molecule and (thus) governs its reactivity.[52] Unfortunately, only a very qualitative statement,[61] based on the perturbation argument, was given about the dependence of x_{min} on the structure of a graph.

A question, which numbers could appear in the spectrum of a graph, can be raised. This problem is far from being solved, though results have been obtained. For example, it is known[28] that only seven Hückel graphs are integral graphs, i.e., the graphs which contain only integers in their spectra.

The existence of the number zero in the spectrum of a graph is of a special importance for the chemical stability of a related molecule, and a number of results are achieved.[26 29,52,165] It is useful to notice that

$$a_N = \det (-\underline{A}) = (-1)^N x_1 x_2 \ldots x_N \equiv (-1)^N \prod_{j=1}^{N} x_j \qquad (2.76)$$

and therefore if the equation

$$a_N = 0 \qquad (2.77)$$

holds, there is, at least, one zero in the spectrum of a graph G.

The properties of Hückel graphs being planar and trichromatic have so far given a very few useful consequences.[132] But, if a Hückel graph is a bichromatic, the situation changes dramatically. Starting from the form (2.53) for the adjacency matrix $\underline{A}$ of such a graph, Coulson and Rushbroke's Pairing Theorem[24] can be proved. The theorem states that if x_j is an eigenvalue of G, $-x_j$ is the eigenvalue of G too, and as its simple consequence eq. (2.76) can be rewritten as

$$a_N = (-1)^{N/2} \prod_{j=1}^{N/2} x_j^2 \qquad (2.78)$$

in the case when N is an even number. Further consequences of the theorem will be discussed later, while the graph-theoretical proof[52] of the pairing theorem, based on the Sachs theorem,[138] will be given in the next paragraph.

2.4.3. Sachs Theorem

The Sachs theorem[138] relates the structure of graph G and the coefficient a_n of the characteristic polynomial $P(G;x)$ of G. $P(G;x)$ was defined by eqs. (2.68) and (2.69) and it can be explicitly written as

$$P(G;x) = \begin{vmatrix} x-A_{11} & -A_{12} & \cdots & -A_{1N} \\ -A_{21} & x-A_{22} & \cdots & -A_{2N} \\ & & & \\ -A_{N1} & -A_{N2} & \cdots & x-A_{NN} \end{vmatrix} = \sum_{j=0}^{N} a_j\, x^{N-j}$$

where N is the number of vertices of G.

Let us determine the a_j's. It is easy to do that for a_N coefficient. If we take x=0, we have

$$P(G;0) = a_N = (-1)^N \det \underline{A}$$

But det$\underline{A}$ depends on the structure of the corresponding graph. In order to prove this let us start from the definition of determinant

$$\det \underline{A} = \sum_{(j_1, j_2, \ldots, j_N)} (-1)^{p(j_1, j_2, \ldots, j_N)}\, A_{1j_1} A_{2j_2} \cdots A_{Nj_N}$$

where the summation goes over all permutations of starting sequence $(1,2,\ldots,N)$, and $p(j_1, j_2, \ldots, j_N)$ is the parity of the particular permutation $(j_1, j_2, \ldots, j_N)$, i.e. the number of inversions in it.

Only the non-vanishing products $A_{1j_1} A_{2j_2} \cdots A_{Nj_N}$ will contribute to det $\underline{A}$. Such products, defined on a graph, were discussed in details in paragraph 2.2.4. and results were summarized by expressions (2.48) and (2.49), showing that set S_N of all Sachs graphs of G needs to be considered in evaluation of det $\underline{A}$. A particular Sachs graph s with N vertices contributes $\pm 2^{r(s)}$ to det $\underline{A}$. As Sachs graphs correspond to some permutations the parity could be ascribed to them. It is easily to show that each edge and ring component with even number of vertices correspond to an odd number of inversions, and each ring component with odd number of vertices to an even number of inversions, thus showing that the parity of a particular Sachs graph s is

$$(-1)^{c_2(s) + c_4(s) + c_6(s) + \ldots}$$

where $c_2(s)$ is the number of edges of s, and $c_4(s)$, $c_6(s)$, ... is the number of even ring components with 4,6,... vertices. Let us denote by $c(s)$ the number of all components of s. It is obvious that the following relations hold

$$c(s) = c_2(s) + c_3(s) + c_4(s) + c_5(s) + \ldots$$

$$N = 2c_2(s) + 3c_3(s) + 4c_4(s) + 5c_5(s) + \ldots =$$

$$= \left[c_3(s) + c_5(s) + \ldots\right] + 2\left[c_2(s) + c_3(s) + 2c_4(s) + 2c_5(s) + \ldots\right]$$

By use of the above relations, it follows:

$$a_N = \sum_s (-1)^{c(s)} \, 2^{r(s)}, \ s\epsilon S_N \tag{2.79a}$$

showing the dependence of a_N coefficient as well as of the determinant of molecular graph on the structure of graph G.

Let us now consider the first derivative of $P(G;x)$. The derivative of the determinant is obtained as the sum of determinants, in the first of them the first column is derivated, in the second the second column is derivated, etc.:

$$P'(G;x) = Nx^{N-1} + \ldots + a_{N-1} =$$

$$= \begin{vmatrix} 1 & -A_{12} & -A_{13} & \cdots & -A_{1N} \\ 0 & x-A_{22} & -A_{23} & \cdots & -A_{2N} \\ 0 & -A_{32} & x-A_{33} & \cdots & -A_{3N} \\ & & & & \\ 0 & -A_{N2} & -A_{N3} & \cdots & x-A_{NN} \end{vmatrix} + \begin{vmatrix} x-A_{11} & 0 & -A_{13} & \cdots & -A_{1N} \\ -A_{21} & 1 & -A_{23} & \cdots & -A_{2N} \\ -A_{31} & 0 & x-A_{33} & \cdots & -A_{3N} \\ & & & & \\ -A_{N1} & 0 & -A_{N3} & \cdots & x-A_{NN} \end{vmatrix} + \ldots$$

For x=0 one has:

$$a_{N-1} = (-1)^{N-1} \left\{ \begin{vmatrix} A_{22} & A_{23} & \cdots & A_{2N} \\ A_{32} & A_{33} & \cdots & A_{3N} \\ & & & \\ A_{N2} & A_{N3} & \cdots & A_{NN} \end{vmatrix} + \begin{vmatrix} A_{11} & A_{13} & \cdots & A_{1N} \\ A_{31} & A_{33} & \cdots & A_{3N} \\ & & & \\ A_{N1} & A_{N3} & \cdots & A_{NN} \end{vmatrix} + \ldots \right\}$$

i.e. a_{N-1} coefficient is the sum of all principal (N-1)th order minors of the adjacency matrix determinant det $\underline{A}$. The first of minors is the determinant det $\underline{A}(G-v_1)$, the second is the determinant det $\underline{A}(G-v_2)$, etc. $G-v_i$ is a subgraph of G obtained by deletion of vertex v_i and its incident edges from graph G. Let us denote by $S_{N-1}(G-v_i)$ the set of all Sachs graphs of $G-v_i$. Then by repeated use of eq. (2.79a) for each minor we have:

$$a_{N-1} = \sum_i \sum_s (-1)^{c(s)} \, 2^{r(s)}, \ s\epsilon S(G-v_i)$$

If we denote by $S_{N-1} \equiv S_{N-1}(G)$ the set of all Sachs graphs with N-1 vertices of G, we can rewrite the previous equation in the form

$$a_{N-1} = \sum_s (-1)^{c(s)} 2^{r(s)}, \quad s \in S_{N-1}$$

The procedure can be continued. It is known from linear algebra that a_j coefficient equals the sum of all principal j th order minors of the matrix determinant det $\underline{A}$ multiplied by factor $(-1)^j$. As the j th order principal minors correspond to the graphs with j vertices obtained by deletion of $(N-j)$ vertices from graph G, we can write using the same argument as before

$$a_j = \sum_s (-1)^{c(s)} 2^{r(s)}, \quad s \in S_j; \quad j=1,2,\ldots,N \qquad (2.79b)$$

$$(a_0 \equiv 1)$$

The symbols in the above formula (<u>Sachs theorem</u>) have the following meaning: s denotes a <u>Sachs graph</u>[52], S_j is a set of all Sachs graphs with j vertices of G, $c(s)$ and $r(s)$ are the total number of components and the total number of cycles (rings), respectively, in a Sachs graph s, and a_j is, of course, a coefficient of the characteristic polynomial we are about to construct. The Sachs graph is defined in paragraph 2.2.4.

By definition $a_0 = 1$. For graphs not containing loops $a_1 = 0$. When S_j is empty set $(S_j = \emptyset)$, then $a_j = 0$.

Formula (2.79b) was originally given by Harary[86] and Sachs[138]. Following the similar way of reasoning the formula for a_j coefficients of the characteristic polynomial for vertex- and edge-weighted graphs was derived[54a]. The extension to Möbius graphs will be explained in section 2.7.

As an example of use of Sachs theorem let us determine the characteristic polynomial of 3,4-dimethylenecyclobutene graph (G). In <u>Fig.2.17.</u> we list all Sachs graphs of G, and the coefficients are directly obtained:

$$a_0 = 1, \; a_1 = 0, \; a_2 = 6(-1)^1 2^0 = -6, \; a_3 = 0,$$

$$a_4 = 7(-1)^2 2^0 + (-1)^1 2^1 = +7-2 = 5, \; a_5 = 0, \; a_6 = (-1)^3 2^0 = -1.$$

The characteristic polynomial of the graph is then given by

$$P(G;x) = x^6 - 6x^4 + 5x^2 - 1$$

with a graph spectrum: 2.25, 0.80, 0.56, −0.56, −0.80, −2.25 .

The advantage of using the Sachs formula is in the independent evaluation of
a single coefficient which may in some cases be the only one needed.

Before presenting the applications of Sachs theorem, we will discuss the equivalence between the topology (the topological eigenvalue problem) and the simple
Hückel molecular orbital (HMO) theory.[100]

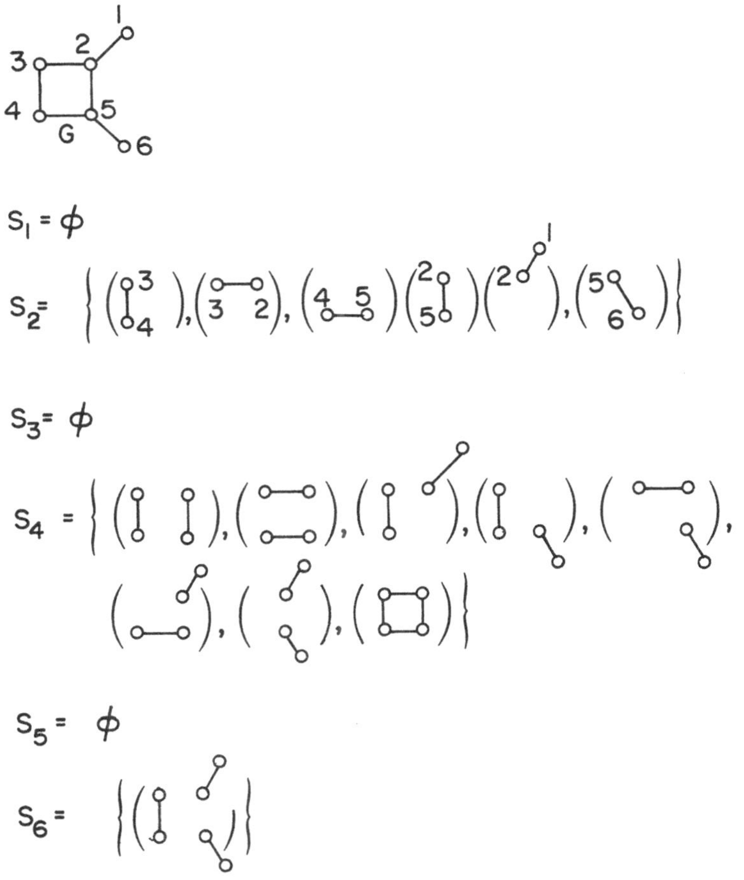

Fig. 2.17. Sachs graphs of the 3,4-dimethylenecyclobutene
molecular graph.

2.5 TOPOLOGY AND SIMPLE MOLECULAR ORBITAL THEORY

The simple molecular orbital theory, named after its originator Hückel,[100,101] has been very often exposed in the literature.[32,121,146]

We assume that the reader is familiar with Hückel theory; therefore, we shall present it here only briefly.

The Hamiltonian matrix $\underline{H}$ in Hückel theory is given by

$$\underline{H} = \alpha\underline{I} + \beta\underline{A} \tag{2.80}$$

where $\underline{I}$ is the unit matrix, $\underline{A}$ an adjacency (topological) matrix of a Hückel graph, α and β are the Coulomb and resonance integrals, respectively, of an effective one-electron Hamiltoninan operator $\hat{H}^{Hückel}$, whose precise structure is not specified, and its matrix elements are given by

$$\langle i|\hat{H}^{Hückel}|i\rangle = \alpha \tag{2.81}$$

$$\langle i|\hat{H}^{Hückel}|j\rangle = \begin{cases} \beta & \text{if atoms i and j are bonded} \\ 0 & \text{otherwise.} \end{cases}$$

The secular determinant of Hückel theory is given in the following form (since the basis functions $|i\rangle$ are orthonormal), assuming $S_{ii}=1$ and $S_{ij}=0$,

$$\det(\underline{H} - E_i\underline{I}) = 0, \quad i = 1,2,...,N. \tag{2.82}$$

Substituting eq. (2.80) for $\underline{H}$ in the above expression one obtains, after some rearrangement

$$\det\left(\frac{E_i-\alpha}{\beta}\underline{I} - \underline{A}\right) = 0. \tag{2.83}$$

If eqs. (2.67) and (2.83) are compared it is seen that the numbers $(E_i-\alpha)/\beta$ are actually making up the spectrum of a Hückel graph,

$$x_i = \frac{E_i-\alpha}{\beta}; \quad i = 1,2,...,N. \tag{2.84}$$

This expression reduces to

$$E_i = x_i \tag{2.85}$$

if β is used as an energy unit and α as the zero energy point. The meaning of the above result is that the <u>eigenvalues of the adjacency matrix are identical with Hückel orbital energy levels.</u>

From eq. (2.80) it is seen that <u>H</u> and <u>A</u> commute and have, therefore, the same eigenvectors. Thus, the <u>eigenvectors of the adjacency matrix are identical with the Hückel molecular orbitals.</u>

Therefore, it is clear that the spectrum of the graph is rather important in Hückel-type calculations. The Hückel theory is in fact fully equivalent to the graph spectral problem[73,78,78a,135,139a,149] However, it must be emphasized that this equivalence is caused by the particular nature of Hückel Hamiltonian in which the short-range forces are dominant[135] Relation (2.80) indicates that Hückel Hamiltonian is a unique function of the adjacency matrix.

$$\underline{H} = \underline{H(A)}. \qquad (2.86)$$

Hence, the topology of a molecule rather than its geometry shapes the form of the simple molecular orbitals.

Hückel (topological) orbitals corresponding to $x_j > 0$, $x_j = 0$ and $x_j < 0$ (see eq. (2.85)) are called <u>bonding</u>, <u>nonbonding</u> and <u>antibonding</u>, respectively. The number of bonding, nonbonding and antibonding orbitals (energy levels) is denoted by N_+, N_0 and N_-, respectively; they are obviously related to the total number of atoms (N) in a conjugated molecule

$$N_+ + N_0 + N_- = N. \qquad (2.87)$$

The total pi-electron energy (Hückel energy) of a conjugated molecule in the ground state is given by

$$E_{pi} = \sum_{j=1}^{N} g_j E_j; \qquad (2.88)$$

or in β units ($\alpha = 0$, $\beta = 1$),

$$E_{pi} = \sum_{j=1}^{N} g_j x_j \qquad (2.89)$$

where g_j is the <u>orbital occupancy number</u>

$$g_j = \begin{cases} 2 & \text{for } x_j > 0 \\ 1 & \text{for } x_j = 0 \\ 0 & \text{for } x_j < 0. \end{cases} \qquad (2.90)$$

E_{pi} is the most important information about a particular conjugated structure. Detailed discussion about E_{pi} is given in Chapter 3, where E_{pi} will be derived in terms of topological parameters of a molecule.

2.6   APPLICATION OF THE COULSON-SACHS GRAPHICAL METHOD

Coulson[21] was the first to propose a graphical method for the evaluation of coefficients of the characteristic polynomial. Elegant formulation of such a procedure
was given in the form of the theorem by Sachs[138] Therefore, we have named this procedure the Coulson-Sachs graphical method[149] This method was described earlier.

The application of the Coulson-Sachs graphical method has produced some general
results for evaluation of a_n values. These can be summarized as follows:

(a) Because there is no Sachs graph possible with a single vertex in graphs associ-
ated with conjugated hydrocarbons, it is always $S_1 = 0$ and $a_1 = 0$. Therefore, since

$$a_1 = \sum_{j=1}^{N} x_j, \tag{2.91}$$

the sum of the whole spectrum is zero,

$$\sum_{j=1}^{N} x_j = 0 \tag{2.92}$$

(b) It can be easily seen that

$$-a_2 = \text{number of bonds} \tag{2.93}$$

$$-\tfrac{1}{3} a_3 = \text{number of three-membered rings.}$$

Note that the following relation holds:

$$a_2 = -\tfrac{1}{2} \sum_{j=1}^{N} x_j{}^2. \tag{2.94}$$

(c) The symbolic expressions for higher coefficients are more complicated. We give
below expressions for a_4, a_5 and a_6 coefficients

$$a_4 = \sum_j \left(\begin{smallmatrix} \circ\!-\!\circ \\ \circ\!-\!\circ \end{smallmatrix} \right)_j - 2 \sum_k \left(n_4 \right)_k \tag{2.95}$$

$$a_5 = \sum_j \left(\triangle \right)_j - 2 \sum_k \left(n_5 \right)_k \tag{2.96}$$

$$a_6 = -\sum_j \left(\begin{smallmatrix} \circ \\ \circ \end{smallmatrix} \right)_j + 2 \sum_k \left(\begin{smallmatrix} \square \end{smallmatrix} \right)_k + 4 \sum_\ell \left(\triangle\triangle \right)_\ell - 2 \sum_m \left(n_6 \right)_m \tag{2.97}$$

n_j (j = 4,5,6) represent four-, five- and six-membered rings. Other symbols are
self-evident. Each graphical combination within braces contributes unity (with an
appropriate sign) to the numerical value of the coefficient. Let us now calculate
directly coefficients of the benzene graph

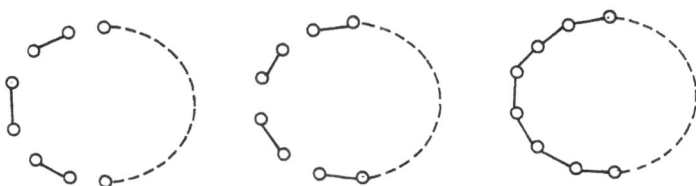

N=6

Following rules for evaluation of a_n coefficients, we obtain
$a_0 = 1$, $a_1 = 0$, $a_2 = -6$, $a_3 = 0$, $a_4 = 9$, $a_5 = 0$ and $a_6 = -4$. General formulae for
a_n were studied by Spialter[143] and Mowshowitz[120]
(d) Since only even-membered cycle subgraphs are possible in bipartite graphs,
$S_n = 0$ and $a_n = 0$ for n odd is always fulfilled for alternant hydrocarbons. Therefore, the characteristic polynomial of alternant hydrocarbon is given by

$$P(G,x) = x^N + a_2 x^{N-2} + a_4 x^{N-4} + \ldots \quad .$$ (2.98)

Thus, the graph spectrum of alternant hydrocarbon must be symmetric with respect to
x = 0. This is yet another proof of the Pairing Theorem of Coulson and Rushbrooke[24]
(e) For annulenes the validity of the Hückel (4m+2) rule can be domonstrated by an-
alysing only a_N coefficients. The Hückel (4m+2) rule states that [4m+2]-annulenes
are thermodynamically more stable than [4m]-annulenes (with the restriction, which
was found later,[32] that the difference decreases for m sufficiently large).

An a_N coefficient for annulenes may be enumerated by constructing all possible
Sachs graphs of a cycle with N vertices. However, in this case there are only three
Sachs graphs possible as shown in Fig. 2.18.

Fig. 2.18. Sachs graphs S_N for annulenes.

Hence, for annulenes

$$a_N = 2(-)^{N/2} 2^0 + (-)^1 2^1 = \begin{cases} 0 & \text{for } N = 4m \\ \\ -4 & \text{for } N = 4m + 2. \end{cases}$$ (2.99)

Therefore, the graph spectrum of [4m]-annulenes must necessarily, because of the pairing theorem, contain two zeros and thus they have triplet ground state (in Hückel theory). [4m+2]-annulenes, of course, have a singlet ground state and are thus more stable than [4m]-annulenes.

(f) In addition, odd-alternant hydrocarbons always have a single zero element in the graph spectrum. This is also a consequence of the pairing theorem. However, it is seen straightforwardly from the application of the Sachs theorem. Consider the graph corresponding to a benzyl radical,

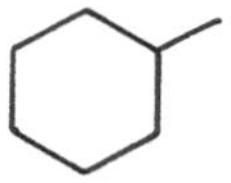

Its characteristic polynomial can be directly written

$$P(G,x) = x^7 - 7x^5 + 13x^3 - 7x \ . \tag{2.100}$$

Since a_N (N = 7) = 0, it follows that one eigenvalue in the spectrum of the benzyl graph is zero.

2.7   EXTENSION OF GRAPH-THEORETICAL CONSIDERATIONS TO MÖBIUS SYSTEMS

Until now, we have treated the pi-electron Hückel systems composed of orbital arrays in which there is no sign inversion among the adjacent 2p pi-orbitals. These systems were represented by Hückel graphs. The sign of 2p pi-orbitals could be changed at the arbitrary chosen atoms, leading to an even number of sign inversions among the adjacent atoms. Obviously, the properties of the system must not be changed under such operation. Therefore, Hückel systems can be defined more generally as those in which there is an even number of sign inversions among the adjacent pi-orbitals.

Let us now introduce the notion of Möbius systems. They are defined[89] as cyclic arrays of orbitals in which there is one sign inversion, or more generally, in which there is an odd number of sign inversions resulting from the negative overlaps between the adjacent pi-orbitals of different sign. Möbius systems can be visualized by use of Möbius strip, which was introduced by Möbius in 1828. Such a strip is obtained if one short edge of a narrow sheet of paper is half-rotated and then joined with the other short edge into a band (see Fig. 2.19). It has only one surface and only one edge. If one short edge is fully rotated and then joined, and if this Möbius strip is bisected,

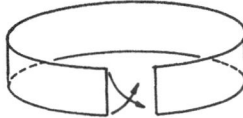 

Fig. 2.19. Möbius strip with one half-twist.

a pair of interlocked rings results. This fact led Frisch and Wasserman[47] to suggest that the interlocked ring molecules could be synthesized starting from Möbius molecules with two half-twists. Some DNA molecules, present in polyoma viruses, also possess Möbius topology[41]

When the pi-orbitals are drawn at a narrow sheet of paper, which is twisted into a Möbius strip (with one half-twist), the Möbius system is represented in a pictorial way. Such systems may appear as transition states in the electrocyclic closures of linear polyenes[161,163] Möbius systems are also called anti-Hückel systems[33] because their stability is governed by rules opposite to those of Hückel.

As the result of twisting around the Mobius strip, the resonance integral (β) between two consecutive pi-orbitals equals the product of the standard β_{CC} with factor $\cos(\pi/N)$, and therefore that β between pi-orbitals at the place where the strip is joined is of opposite sign in comparison with the standard β_{CC} value. Approximately, the whole effect of the consecutive twisting of pi-orbitals can be assigned only to a pair of orbitals of different sign. As a result, the resonance integral, corresponding to such a pair, will only change the sign in comparison to the standard β_{CC} value. We also assume that Hückel MO theory can be applied to Möbius systems.

In order to extend the graph-theoretical considerations to Möbius systems, we need to define a new type of graphs which are called Möbius graphs[53,54] and are denoted by G_M. (There should be no confusion with the Möbius ladder graphs?) The weight of edges in Möbius graphs is either +1 or -1 depending on whether two (adjacent) pi-orbitals in a molecule are in the positive-positive (+1) or in the positive-negative (-1) overlap relationships, respectively.

As an example, Fig. 2.20. shows graphs G_1 and G_2 which correspond to Hückel and Möbius cyclobutadiene, respectively.

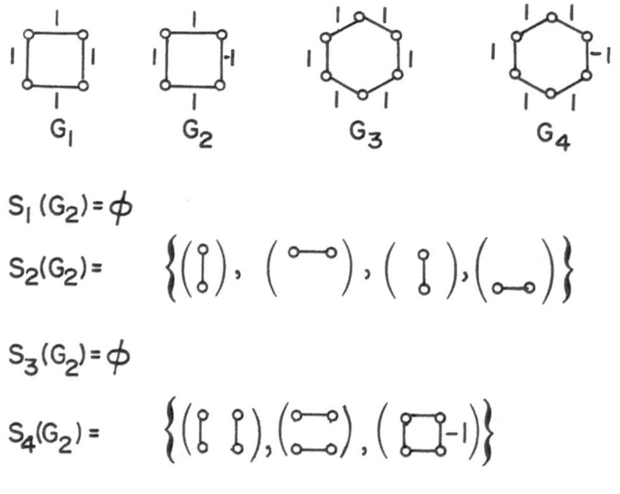

$$S_1(G_2) = \phi$$

$$S_2(G_2) = \left\{ \left(\begin{smallmatrix} \circ \\ \circ \end{smallmatrix}\right), \left(\circ\!-\!\circ\right), \left(\begin{smallmatrix} \circ \\ \circ \end{smallmatrix}\right), \left(\circ\!-\!\circ\right) \right\}$$

$$S_3(G_2) = \phi$$

$$S_4(G_2) = \left\{ \left(\begin{smallmatrix} \circ & \circ \\ \circ & \circ \end{smallmatrix}\right), \left(\circ\!-\!\circ \atop \circ\!-\!\circ\right), \left(\Box\, \text{-1}\right) \right\}$$

Fig. 2.20. Hückel and Möbius graphs.

The location of the edge with the weight -1 in the Möbius graphs of annulenes is arbitrary, but the important information is given by stating that one (odd) -1 is present.

The Möbius graphs are a special case of more general graphs with an arbitrary number of -1 edges which we call the generalized graphs. Generalized graphs occur, for example, when the graph theory is applied to systems treating the relationships between people, thus, +1 edge would symbolize the attraction between two persons, 0 indifference, and -1 edge, repulsion. In other words, in the case of generalized graphs there are two distinctive binary relations R_+ and R_- of opposite meaning, defined on the set of vertices of the graph.

The adjacency matrix $\underline{A}$ assigned to the Möbius graph G_M is defined as follows:

$$A_{ij} = \begin{cases} +1 & \text{if } (i,j)\epsilon R_+ \\ -1 & \text{if } (i,j)\epsilon R_- \\ 0 & \text{otherwise} \end{cases} \tag{2.101}$$

or alternatively,

$$A_{ij} = \begin{cases} +1 & \text{if there is a positive edge between the adjacent vertices } v_i \text{ and } v_j \\ -1 & \text{if there is a negative edge between the adjacent vertices } v_i \text{ and } v_j \\ 0 & \text{otherwise.} \end{cases} \tag{2.102}$$

As example, the adjacency matrices $\underline{A}(G_1)$ and $\underline{A}(G_2)$ of Hückel (G_1) and Möbius (G_2) cyclobutadiene graphs in $\underline{\text{Fig. 2.20.}}$ are

$$\underline{A}(G_1) = \begin{bmatrix} 0 & 1 & 0 & 1 \\ 1 & 0 & 1 & 0 \\ 0 & 1 & 0 & 1 \\ 1 & 0 & 1 & 0 \end{bmatrix} \qquad \underline{A}(G_2) = \begin{bmatrix} 0 & 1 & 0 & 1 \\ 1 & 0 & -1 & 0 \\ 0 & -1 & 0 & 1 \\ 1 & 0 & 1 & 0 \end{bmatrix}. \tag{2.103}$$

In the same way as for the Hückel graphs, the Hamiltonian matrix $\underline{H}$ and the adjacency matrix $\underline{A}$, corresponding to Möbius system, can be made identical, thus showing that the molecular topology, in the framework of simple MO-theory, will influence the chemical properties of Möbius systems, too.

First, we shall see how the structure of the Möbius graph (G_M) is related to the coefficients (a_j's) appearing in the characteristic polynomial $P(G_M;x)$ of the adjacency matrix $\underline{A}(G_M)$. The construction of the characteristic polynomial, appropriate to a Möbius graph, is once again a purely combinatorial problem and may be carried out in the same way as described before for Hückel graphs using the Sachs formula.

Since the matrices $\underline{A}(G_M)$ are formally a special case of adjacency matrices for edge-weighted graphs, i.e., a special case: $h = 0$, $k = +1$ or -1, the considerations which led to formula (2.79) can be applied. Because for an edge component of a Sachs graph s, the corresponding k was squared, it follows that such a component contributes factor +1, regardless of the sign of that edge ($k = +1$ or -1). For a ring component

of a Sachs graph s, the corresponding k's were multiplied "around" the ring, and therefore such a component contributes +1 if the number of -1 edges in the ring component
is even and -1 if this number is odd. Therefore the following extended Sachs formula,
which covers both Hückel (G) and Möbius (G_M) graphs is valid:

$$a = 1 \text{ (per definitionem)}$$
$$a_j = \sum_s (-1)^{c(s)+p_{r(s)}} \; 2^{r(s)} (s \in S_j) \text{ for } 1 \leqslant j \leqslant N \tag{2.104}$$

where the symbols have the same meaning as in eq. (2.79) while $p_{r(s)}$ is the number of
-1 edges in the ring components of Sachs graphs. Let us apply the above formula to
Möbius cyclobutadiene graph G_2 in Fig. 2.20. All Sachs graphs of G_2 are shown in the
same figure. The use of the above formula gives

$$a_0(G_1) = 1$$
$$a_1(G_2) = 0$$
$$a_2(G_2) = 4(-1)^{1+0}2^0 = -4 \tag{2.105}$$
$$a_3(G_2) = 0$$
$$a_4(G_2) = 2(-1)^{2+0}2^0 + (-1)^{1+1}2^1 = 4$$

and

$$P(G_2;x) = x^4 - 4x^2 + 4. \tag{2.106}$$

The polynomial (2.106) could be compared with the characteristic polynomial belonging to Hückel cyclobutadiene (G_1),

$$P(G_1;x) = x^4 - 4x^2. \tag{2.107}$$

Obviously, polynomials $P(G_1;x)$ and $P(G_2;x)$ differ only in the value of the a_N coefficient.

A similar result can be also obtained for Hückel (G_3 in Fig. 2.20.) and Möbius
(G_4) benzene graphs,

$$P(G_3;x) = x^6 - 6x^4 + 9x^2 - 4$$
$$P(G_4;x) = x^6 - 6x^4 + 9x^2,$$

but now the a_N coefficient for the Möbius graph vanishes, while for the Hückel graph
it differs from zero.

Indeed, because monocyclic systems contain the ring components only in S_N set of
Sachs graphs, it follows the general characteristic feature of all monocyclic Hückel
and Möbius systems (annulenes) that they differ in the value of the a_N coefficient

only.

The Sachs graphs of S_N for Möbius annulenes are shown in <u>Fig. 2.21.</u> and from them can be read

$$a_N = 2(-1)^{N/2} \, 2^0 + (-1)^{1+1} \, 2^1 = \begin{cases} 4 & N = 4m \\ \\ 0 & N = 4m+2. \end{cases} \qquad (2.109)$$

<u>Fig. 2.21.</u> Sachs graphs S_N for Möbius annulenes.

The a_N coefficient provides information about the chemical stability, as described in Section 2.6. The results of eqs. (2.99) and (2.109), obtained for Hückel and Möbius cyclobutadienes and benzenes, can be summarized in a very simple classification scheme of all $4m+2 \equiv 2 \pmod 4$ and $4m \equiv 0 \pmod 4$ generalized systems, i.e., Hückel and Möbius annulenes:

$$a_N = \mp 4 \quad \begin{cases} \text{Hückel } (4m+2)\text{-systems} \\ \\ \text{Möbius } (4m)\text{-systems} \end{cases}$$

and
(2.110)

$$a_N = 0 \quad \begin{cases} \text{Hückel } (4m)\text{-systems} \\ \\ \text{Möbius } (4m+2)\text{-systems} \end{cases}$$

3.   TOTAL pi-ELECTRON ENERGY

3.1.   INTRODUCTION

Total pi-electron energy (E_{pi}) is one of the most important pieces of information about a conjugated molecule which can be obtained from simple Hückel molecular orbital (HMO) calculations.   Moreover, HMO E_{pi}'s are often of the same degree of quantitative accuracy as those calculated by much more sophisticated SCF MO models.   An intriguing result along these lines was obtained by Schaad and Hess[139] who showed that in many cases E_{pi} follows linearly the total (thermodynamically measurable) energy of a conjugated compound.   The physical reasons of how it is possible that such a simple model like HMO can give not only qualitative but sometimes also fair quantitative agreement with experimental findings are at present not well understood.   However, in this chapter we shall not be involved in clarification of this complicated question, but we proceed with an "empirical" assumption that HMO theory works and that it is expecially reliable in reproducing total pi-electron energies.   <u>Our considerations will be always on a qualitative level.</u>

Hence, the question to be posed is how E_{pi} depends on the structure of a conjugated compound.   Since the HMO model is purely topological, it follows that E_{pi} (as calculated within the HMO approximation) is a function of the molecular topology only, that it (in the terminology of the present notes) depends on the structure of the molecular graph.*   According to the authors' opinion, the answer to this problem is at least of the same importance as the ability of producing numbers by quantum-chemical calculations.

The investigation of general properties of E_{pi} has a relatively long history. The first important paper on this problem was published by Coulson[20] in 1940, although partial results have been obtained even earlier[101]   In the last three decades there has been a continuous interest in total pi-electron energy, and this problem has been attacked from various viewpoints and by various mathematical and/or conceptual methods[3,20,22,23,42,81-83,99,109,118,119,136,145,158]   Almost this whole

---

*Of course, it would be a serious error to consider E_{pi} as depending solely on molecular topology. E_{pi} (as well as any physicochemical property of a compound) is sensitive to all geometric (conformational) details of the molecule.   In particular, E_{pi} depends essentially on whether a conjugated system is planar or not.   It is assumed throughout this book that the reader is aware of the fact that <u>we are discussing only one aspect of molecular reactivity.</u>   Thus, in the interpretation of concrete chemical findings the "nontopological" structural factors need also to be taken into account.

Fortunately, some pi-electron properties, and among them especially E_{pi}, are rather sensitive to topological factors and relatively insensitive to molecular conformation. The mere fact that HMO theory works is the best argument in favour of this assumption.

chapter will deal with the exposure of algebraic properties of E_{pi}, while applications are mainly contained in the subsequent chapter. The interested reader will see which fascinating chemical rules can be deduced and/or proved by combining the matrix calculus, algebra of polynomials, functional analysis and, of course, the proof techniques of graph theory.

Those readers who are not inclined towards mathematical speculations are advised to go immediately to the summary of this chapter, as well as to the first sections of the subsequent chapter.

In the following we shall express E_{pi} in β units ($\alpha = 0$, $\beta = 1$). Therefore E_{pi} is always positive and larger E_{pi} means larger thermodynamical stability. Now,

$$E_{pi} = \sum_{j=1}^{N} g_j \, x_j \tag{3.1}$$

where g_j is the occupation number of the jth MO. Let the considered molecule have N conjugated atoms and N_e pi-electrons. Then the ground state total pi-electron energy is given by

$$E_{pi} = \begin{cases} 2 \sum_{j=1}^{N_e/2} x_j & \text{if } N_e = \text{even} \\[2ex] 2 \sum_{j=1}^{(N_e-1)/2} x_j + x_{(N_e+1)/2} & \text{if } N_e = \text{odd}. \end{cases} \tag{3.2}$$

In the great majority of cases considered in this book, $N_e = N$ and therefore

$$E_{pi} = \begin{cases} 2 \sum_{j=1}^{N/2} x_j & \text{if } N = \text{even} \\[2ex] 2 \sum_{j=1}^{(N-1)/2} x_j + x_{(N+1)/2} & \text{if } N = \text{odd}. \end{cases} \tag{3.3}$$

3.2   IDENTITIES AND INEQUALITIES

3.2.1.   The Fundamental Identity

Let there be N_+, N_0 and N_- positive, zero and negative elements in the spectrum of the molecular graph. Define W_+ and W_- as the sum of the positive and negative elements of the spectrum

$$W_+ = \sum_{j=1}^{N_+} x_j \quad \text{and} \quad W_- = \sum_{j=N_++N_0+1}^{N_-} x_j . \tag{3.3a}$$

Of course,

$$W_+ + W_- = 0 \tag{3.4}$$

and

$$W_+ - W_- = \sum_{j=1}^{N} |x_j| . \tag{3.5}$$

Now assume that all the N_+ bonding levels in the molecule are doubly occupied and all the N_- antibonding levels are unoccupied. Whether there are pi-electrons in the NBMOs or not has no influence on the value of E_{pi}. Then from eqs. (3.2) or (3.3) it follows

$$E_{pi} = 2W_+ . \tag{3.6}$$

Because of eq. (3.4), $2W_+ = W_+ - W_-$ and we obtain the identity

$$E_{pi} = \sum_{j-1}^{N} |x_j| . \tag{3.7}$$

We emphasize once again that relation (3.7) holds exactly in the case of filled bonding and empty antibonding MOs.

Note that relation (3.7) is valid for all alternant molecules because of the pairing theorem. In fact, the identify is true if $N_+ = N_-$.

Now suppose $N_+ \neq N_-$ and there is a filled antibonding level $x_{N/2} < 0$. Then

$$E_{pi} = \sum_{j=1}^{N} |x_j| + 2x_{N/2} . \tag{3.8}$$

If the conjugated system is not a very small one, $\Sigma|x_j| \gg 2x_{N/2}$ and relation (3.7) holds as a good approximation. Similarly, if there is present an empty bonding level $x_{N/2+1} > 0$,

$$E_{pi} = \sum_{j=1}^{N} |x_j| - 2x_{N/2+1} \tag{3.9}$$

and since $\Sigma|x_j| >> 2x_{N/2+1}$, relation (3.7) is again fulfilled approximately.

Thus, we have demonstrated that relation (3.7) holds either exactly or as a good approximation for all molecular graphs. In the following we will refer to this result as the <u>fundamental identity</u> for E_{pi}. It will be the basis for all further considerations.

3.2.2   <u>Relations between E_{pi}. The Adjacency Matrix and the Density Matrix</u>

The eigenvalue equation of the adjacency matrix $\underline{A}$ is

$$\underline{C}_j \, \underline{A} = x_j \, \underline{C}_j \quad (j = 1,2,\ldots,N) \tag{3.10}$$

which can be written also as

$$\underline{C} \, \underline{A} = \underline{X} \, \underline{C} \tag{3.11}$$

where

$$\underline{C} = \begin{bmatrix} \underline{C}_1 \\ \underline{C}_2 \\ \cdot \\ \cdot \\ \cdot \\ \underline{C}_N \end{bmatrix} \tag{3.12}$$

and*

$$\underline{X} = \text{diag} \, (x_1,x_2\ldots,x_N). \tag{3.13}$$

Then $\underline{C}$ is a unitary matrix with the properties $\underline{C} \, \underline{C}^+ = \underline{C}^+ \, \underline{C} = \underline{I}$. Let the function $f(x)$ be defined for all x_j (j=1,2,\ldots,N). Let $f(\underline{X}) = \text{diag} \, (f(x_1),f(x_2),\ldots,f(x_N))$. Then, by definition, that matrix function $f(\underline{A})$ is given by

$$\underline{C} \, f \, (\underline{A}) = f \, (\underline{X}) \, \underline{C} \tag{3.14}$$

or

$$f \, (\underline{A}) = \underline{C}^+ \, f \, (\underline{X}) \, \underline{C} \tag{3.15}$$

or in scalar notation

$$[f(\underline{A})]_{pq} = \sum_{j=1}^{N} f(x_j) \, C_{jp} \, C_{jq}. \tag{3.16}$$

This formula enables one[136],[137] to find useful relations between $\underline{A}$, E_{pi} and the

---

*$\underline{M}=\text{diag}(M_1,M_2,\ldots,M_N)$ denotes a diagonal matrix, the matrix elements of which are $M_{jj} = M_j$ and $M_{ij}=0$ for $i \neq j$.

density matrix $\underline{P}$ defined as

$$P_{pq} = \sum_{j=1}^{N} g_j \, C_{jp} \, C_{jq}. \tag{3.17}$$

If we introduce a formal function $g=g(x)$ such that $g(x_j) = g_j$, we obtain from eq. (3.16)

$$\underline{P} = g(\underline{A}). \tag{3.17a}$$

In the case of filled bonding and empty antibonding MOs,

$$x_j \, g(x_j) = x_j + |x_j| \tag{3.18}$$

from which it follows immediately

$$\underline{A}\,\underline{P} = A + |\underline{A}|. \tag{3.19}$$

Note that $|\underline{A}|$ denotes the absolute value of the matrix $\underline{A}$ defined as

$$|A| = \underline{C}^{+} \, \text{diag} \, (|x_1|, |x_2|, \ldots, |x_N|)\underline{C}. \tag{3.20}$$

Since the trace Tr of a matrix* is not changed by a unitary transformation,

$$\text{Tr} \, |\underline{A}| = \text{Tr diag} \, (|x_1|, |x_2|, \ldots, |x_N|) = \sum_{j=1}^{N} |x_j| \tag{3.21}$$

and because of the fundamental identity,

$$E_{pi} = \text{Tr} \, |\underline{A}|. \tag{3.22}$$

Another equality for E_{pi} is obtained from eq. (3.19), taking into account that $\text{Tr} \, \underline{A} = 0$, namely

$$E_{pi} = \text{Tr} \, \underline{A} \, \underline{P}. \tag{3.23}$$

Both identities (3.22) and (3.23) were first obtained by Ruedenberg[137] although a special case of them was known earlier to Hall[82]

---

*The trace of a matrix $\underline{M}$ is the sum of its diagonal elements, $\text{Tr} \, \underline{M} = \Sigma_j \, M_{jj}$.

3.2.3. The Loop Rule

In the preceding section it was demonstrated that the problem of finding E_{pi} is reduced to the problem of finding the trace of a matrix. On the other hand we know that

$$\text{Tr } \underline{A} = 0$$
$$\text{Tr } \underline{A}^2 = 2M$$
$$\text{Tr } \underline{A}_3 = 6 \text{ x number of triangles in the molecular graph.}$$

These relations are generalized by the following theorem:

$$\text{Tr } A^n = L_n = \text{number of loops of length n in the graph.}$$

We call[70] a loop (of length n) a closed walk (of length n) in the graph. Note that every cycle is a loop, but every loop is not a cycle.

This all suggests[67,70,82,137] trying to express $\text{Tr } |A|$ in terms of $\text{Tr } \underline{A}^n$'s.
Unfortunately, there exists no expansion of the form $|A| = \sum_j u_j \underline{A}^{2j}$ because of the
nonanalytical behaviour of $|x|$ for $x = 0$.

Let us approximate $\underline{A}$ by a finite matrix polynomial $\sum_{j=0}^{t} u_j \underline{A}^{2j}$. Then,

$$E_\pi \approx \sum_{j=0}^{t} u_j L_{2j}. \tag{3.24}$$

It can be proved[70] that independently of the value of t, $u_0 > 0$, $u_1 > 0$, $u_2 < 0$,
$u_3 > 0$, $u_4 < 0$, $u_5 > 0$,... In other words $u_{2j} < 0$ and loops of the length 4 j decrease E_{pi}; $u_{2j+1} > 0$ and loops of the length 4j+2 increase E_{pi}. This conclusion is
named the loop rule. It shows that E_{pi} is dependent on loops contained in the graph
and exhibits the differences between (4j)- and (4j+2)-membered loops. Evidently, the
loop rule resembles the famous Hückel (4j+2) rule.

Using the loop rule and taking different values for t (t=0,1,2,3), approximate
formulae for E_{pi} could be deduced[67,82,137] They enabled for the first time an insight
into the dependence of E_{pi} on graph structure. However, these formulae suffer from
containing a number of loosely defined constants (u_j) and are rather complicated. We
will not give them here.

Further loop rule-type results are given in Section 3.6.2.

3.2.4. Inequalities for E_{pi}

Total pi-electron energy is obviously a bounded quantity. Namely, if the maximal
vertex degree in the graph is d_{max},

$$-d_{max} \leqslant x_j \leqslant d_{max} \tag{3.25}$$

and therefore

$$0 \leqslant E_{pi} \leqslant N \ d_{max}. \tag{3.26}$$

McClelland[119] was the first to show that much better bounds for E_{pi} can be deduced. Let

$$F = 2M - N \ (\det \underline{A})^{2/N}. \tag{3.27}$$

Note that $F \geqslant 0$. Then, McClelland's inequalities are

$$0 \leqslant 2NM - E_{pi}^2 \leqslant (N-1) \ F. \tag{3.28}$$

These relations were later further improved[58]

$$F \leqslant 2NM - E_{pi}^2 \leqslant (N-1) \ F \tag{3.29}$$

which hold for all graphs

$$F \leqslant 2NM - E_{pi}^2 \leqslant (N-2) \ F \tag{3.30}$$

which hold for bipartite graphs.

The importance of these inequalities is in that they show which are the three most important topological factors in determining E_{pi}: the number of atoms (N), bonds (M) and the algebraic structure count (det $\underline{A}$).

McClelland has concluded that from $E_{pi} \leqslant \sqrt{2MN}$, an approximate formula

$$E_{pi} \approx a\sqrt{2MN}, \quad a = \text{const} \tag{3.31}$$

can be obtained by least-squares fitting of a. The optimal value $a = 0.92$ reproduced E_{pi} up to 95%.

In Section 3.6.2. it will be shown that in case of $N_0 > 0$ in all eqs. (3.27)-(3.31) N can be replaced by $\tilde{N}$ and det $\underline{A}$ by Δ, where

$$\tilde{N} = N - N_0 \tag{3.32}$$

and Δ is the product of the absolute values of all nonzero elements in the graph spectrum.

$$\Delta = (-1)^{N-} (\prod_{j=1}^{N_+} x_j) \ (\prod_{j=N_+ + N_0 + 1}^{N} x_j). \tag{3.33}$$

Of course, if there are no zeros in the graph spectrum, $\tilde{N} = N$ and $\Delta = |\det \underline{A}|$.

3.3   THE COULSON INTEGRAL FORMULA

Another class of identities between E_{pi} and the characteristic polynomial of the graph is initiated by the classical work of Coulson[20] These identities play a crucial role in recent topological studies of E_{pi}. Therefore, we will consider them in detail.

3.3.1.   The First Integral Formula

When an HMO problem is solved by hand calculation, one first determines the characteristic polynomial $P(G;x)$ of the molecular graph

$$P(G;x) = \det(x\underline{I} - \underline{A}) \tag{3.34}$$

which is expressed in the form

$$P(G;x) = \sum_{j=0}^{N} a_j \, x^{N-j}. \tag{3.35}$$

The next step is the solution of the equation $P(G;x) = 0$ and the evaluation of the roots x_j of the characteristic polynomial (=graph spectrum). From x_j's, E_{pi} is calculated immediately. It is natural to ask about a possibility to compute E_{pi} directly from $P(G;x)$ without the knowledge of its roots.

Coulson has offered[20] an elegant solution of this problem, namely

$$E_{pi} = \frac{1}{\pi} \int_{-\infty}^{+\infty} [N - ix \frac{P'(G;ix)}{P(G;ix)}]dx \tag{3.36}$$

where $i = \sqrt{-1}$. Instead of using contour integration and complex function analysis (as in Ref. 20), we derive Coulson's formula in an elementary way[51]

Since

$$P(G;x) = \prod_{j=1}^{N} (x-x_j) \tag{3.37}$$

we have

$$\frac{P'(G;x)}{P(G;x)} = \sum_{j=1}^{N} (x-x_j)^{-1} \tag{3.38}$$

where $P'(G;x)$ is the first derivative of $P(G;x)$. Furthermore, in order to simplify the notation, we will use the abbreviation

$$\frac{1}{\pi} \int_{-\infty}^{+\infty} F(x) \ dx \equiv \ < F(x) > \ . \tag{3.39}$$

Thus, consider the elementary integrals

$$I_1 = <t^2/(t^2+x^2)> = |t| \tag{3.40}$$

$$I_2 = <tx/(t^2+x^2)> = 0. \tag{3.41}$$

Then,

$$|t| = I_1 + iI_2 = <t(t+ix)/(t^2+x^2)> = <t/(t-ix)> \tag{3.42}$$

$$= <1 - ix/ (ix-t)> \ .$$

When this expression is substituted back into the fundamental identity (3.7), we obtain the relation

$$E_{pi} = \sum_{j=1}^{N} <1 - ix/(ix - x_j)>$$
$$= <N - ix \sum_{j=1}^{N} (ix-x_j)^{-1}> \tag{3.43}$$

which combined with eq. (3.38) gives

$$E_{pi} = <N-ixP'(G;ix)/P(G;x)> \ . \tag{3.44}$$

This is exactly the Coulson formula (3.36) written in another form.

Eqs. (3.36) and (3.44) give a relation between E_{pi} and the characteristic polynomial, but not between E_{pi} and the graph structure. A long time after this formula has been obtained, it was not known how $P(G;x)$ depends on the structure of a graph. This seems to be the reason why Coulson's integral formula (3.36) was usually treated in quantum chemistry textbooks and monographs as a mathematical curiosity and had no impact on investigations of E_{pi} for decades.

This situation was substantially changed after the Sachs theorem had been applied to HMO theory[52]. Almost simultaneously Hosoya discovered[98] the same result in another form. Thereafter it became clear how the structure of a graph is reflected in $P(G;x)$. By combining the Coulson-type formulae with the Sachs theorem one has in principle a complete insight into the dependence of E_{pi} on molecular topology.

In 1965 Marcus[118] was the first to use an integral formula for E_{pi} in combination with graph theory.

3.3.2.   Further Coulson-Type Formulae. I.

Eq. (3.44) can be written also as

$$E_{pi} = \langle N - x\frac{d}{dx} \ln P(G;ix)\rangle .$$ (3.45)

However, if there are zeros in the graph spectrum, $\ln P(G;ix)$ is infinite for $x = 0$. Therefore, it is useful to introduce the polynomial $\tilde{P}(G;x)$

$$\tilde{P}(G;x) = x^{-N_0} P(G;x) = \sum_{j=0}^{\tilde{N}} a_j x^{\tilde{N}-j} .$$ (3.46)

Then from eq. (3.45),

$$E_{pi} = \langle \tilde{N} - x\frac{d}{dx}\ln \tilde{P}(G;ix)\rangle .$$ (3.47)

The reader should remember that $\tilde{N} = N - N_0$. Formula (3.47) can be formally obtained if $\tilde{N}$ is substituted instead of N in eq. (3.45). This should be evident since $P(G;x)$ has the same roots as $\tilde{P}(G;x)$ except for the zero roots. But nonbonding pi-electrons do not influence the value of E_{pi}.

Therefore, we conclude that in the case of $N_0 > 0$ one can substitute $\tilde{N}$ instead of N in any topological formula for E_{pi}.

Substituting $x' = 1/x$ into eq. (3.44) and performing convenient transformations, we obtain a formula[52]

$$E_{pi} = \langle x^{-2} \ln |H(G;x)|\rangle$$ (3.48)

where

$$H(G;x) = (ix)^N P(G;-i/x) .$$ (3.49)

In the general case the polynomial $H(G;x)$ is complex. However, for bipartite graphs, $H(G;x)$ reduces to

$$H(G;x) = \sum_{j=0}^{\tilde{N}/2} |a_{2j}| x^{2j} .$$ (3.50)

This finally gives

$$E_{pi} = \langle x^{-2} \ln \sum_{j=0}^{\tilde{N}/2} |a_{2j}| x^{2j}\rangle .$$ (3.51)

An important consequence of this formula is that in bipartite graphs E_{pi} is a monotonously increasing function of all $|a_{2j}|$'s.

In non-bipartite graphs the explicit form of eq. (3.48) is

$$E_{pi} = \tfrac{1}{2} < x^{-2} \ln(U^2 + x^2 V^2) > \qquad (3.52)$$

with

$$U = U(x) = \sum_j (-1)^j a_{2j} x^{2j} \qquad (3.53)$$

and

$$V = V(x) = \sum_j (-1)^j a_{2j+1} x^{2j} . \qquad (3.54)$$

3.3.3   An Application of the Coulson Integral Formula: The Tree with
        Maximal Energy

We prove here that among acyclic graphs (=trees) the path P_N has the maximal energy. The path is, of course, the graph representation of the linear polyene (see Fig. 3.1.).

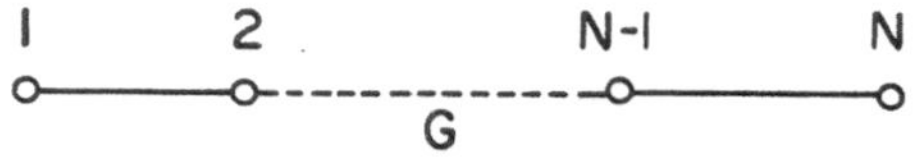

Fig. 3.1. The path P_N.

Every tree has terminal vertices(i.e., vertices of degree one). If p is a vertex and q is its neighbour,

$$P(G;x) = x P(G-p;x) - P(G-p-q;x) \qquad (3.55)$$

which is a special case of the Heilbronner formula[87] (The reader should remember that G-p and G-p-q are graphs obtained by deletion of the vertex p and the vertices p and q from a graph G, respectively.)

Since any acyclic graph is bipartite, eq. (3.51) applies. Moreover, from eq. (3.55) we deduce

$$\left| a_{2j}(G) \right| = \left| a_{2j}(G-p) \right| + \left| a_{2j-2}(G-p-q) \right|. \qquad (3.56)$$

Note that G-p has N-1 and G-p-q N-2 vertices.

Now we can prove the following theorem:
Theorem: Among trees with N vertices, the path P_N has maximal energy. Proof will be performed by induction. For N=2,3,4 the theorem can be verified by checking all the possible cases. Assume the theorem is valid for all trees with 2,3,...,N-1 vertices. We prove that it holds for trees with N vertices too.

From eq. (3.51) it follows that $E_{pi}(G)$ will be maximal if all $\left| a_{2j}(G) \right|$'s are maximal. Further, from eq. (3.56) it follows that $|a_{2j}(G)|$ will be maximal if both $|a_{2j}(G-p)|$ and $|a_{2j-2}(G-p-q)|$ are maximal. According to our assumption, this will occur if $G-p = P_{N-1}$ and $G-p-q = P_{N-2}$. Then necessarily $G = P_N$. Q.E.D.

It can also be proved that the acyclic polyene with the second maximal E_{pi} is the 3-vinyl derivative of the linear polyene, the molecular graph of which is H in Fig. 3.2.

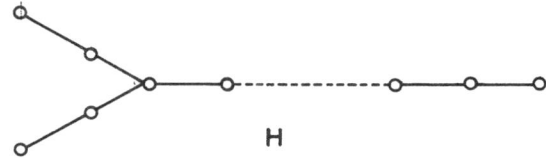

$$H$$

Fig. 3.2. The acyclic polyene with the second maximal E_{pi}.

3.3.4.   Further Coulson-Type Formulae.   II.

Let G_1 and G_2 be two molecular graphs with N_1 and N_2 vertices and let the corresponding energies be $E_{pi}(G_1)$ and $E_{pi}(G_2)$. From eq. (3.47) we have

$$E_{pi}(G_1) - E_{pi}(G_2) = <\tilde{N}_1 - \tilde{N}_2 - x \frac{d}{dx} \ln \frac{\tilde{P}(G_1;ix)}{\tilde{P}(G_2;ix)} >. \qquad (3.57)$$

If $\tilde{N}_1 = \tilde{N}_2$, it can be proved[22,118] that

$$E_{pi}(G_1) - E_{pi}(G_2) = <\ln|\frac{\tilde{P}(G_1;ix)}{\tilde{P}(G_2;ix)}|> . \qquad (3.58)$$

Special cases of this formula are obtained[22] for $\tilde{P}(G_1;x) = (x^2-1)^{\tilde{N}/2}$ and $\tilde{P}(G_2;x) = (x^2-2M/\tilde{N})^{\tilde{N}/2}$:

$$E_{pi} - \tilde{N} = <\ln|\tilde{P}(G;ix)|/(x^2 + 1)^{\tilde{N}/2}> \qquad (3.59)$$

$$E_{pi} - \sqrt{2M\tilde{N}} = <\ln|\tilde{P}(G;ix)|/(x^2 + 2 M/\tilde{N})^{\tilde{N}/2}> . \qquad (3.60)$$

Formula (3.60) expresses in fact the energy difference between isomers, while formula (3.59) is an integral expression for classical resonance energy. We easily recognize the McClelland-type term $\sqrt{2M\tilde{N}}$ in eq. (3.60). The right side of this equality can be thus understood as a nonempirical correction to formula (3.31).

3.3.5.   A Class of Approximate Topological Formulas for E_{pi}.

Recently a technique has been developed[60a] which enables one to estimate the integrals

(3.44),(3.45),(3.47),(3.48), etc. In this section we will show first that a formula similar to eq. (3.31) can be obtained from eq. (3.44). Let us consider the function

$$F(x) = N-ix \, P'(G;ix)/P(G;ix).$$ (3.61)

Analysis shows[51,60a] the $F(x)$ is a bell-shaped function with $F_{max} = F(0) = \tilde{N}$ and $F(x) \sim 2M/x^2$ for large x. The simplest algebraic function having these properties is

$$F^*(x) = 2M\tilde{N} - (2M + \tilde{N}x^2)^{-1}$$ (3.62)

which yeilds immediately the McClelland-type expression

$$E_{pi} \approx <F^*(x)> = \sqrt{2M\tilde{N}} \, .$$ (3.63)

The above integration may be performed without difficulties.

Thus, using a correct model function $F^*(x)$ instead of the exact function $F(x)$, the integration can be performed and an approximate topological formula is obtained.

When the same procedure is applied to other Coulson integral formulae, a whole class of topological expressions for E_{pi} is gained. Hence the integral (3.48) can be replaced by

$$E_{pi} \approx <[MT^2 + (\tilde{N}/2) \ln(1 + x^2)](T^2 + x^2)^{-1}>$$ (3.64)

which yields after straightforward integration

$$E_{pi} \approx MT + \tilde{N} \, \frac{\ln(1 + T)}{T}$$ (3.65)

with T being a parameter which can be determined as

$$T = \sqrt{\frac{2M - \tilde{N} \ln 2}{2M - 2\ln Z}}$$ (3.66)

where

$$Z = \sum_{j=0}^{\tilde{N}/2} |a_{2j}| \, .$$ (3.67)

The topological index Z was introduced and discussed in detail by Hosoya.[97-99] There was also shown that a relation

$$E_{pi} \approx a \ln Z, \quad a = const$$ (3.68)

holds for certain classes of conjugated systems.

Approximating formulae (3.59) and (3.60) one obtains

$$E_{pi} \approx \tilde{N} + \sqrt{(M - \tilde{N}/2)\ \ln \Delta} \tag{3.69}$$

$$E_{pi} \approx \sqrt{2M\tilde{N}} - \tfrac{1}{2}\sqrt[4]{XY^3} \tag{3.70}$$

where

$$X = \tilde{N}/2\ \ln(2M/\tilde{N}) - \ln \Delta \tag{3.71}$$

$$Y = D/2 + 2n_4 - M^2/\tilde{N} - M/2 \tag{3.72}$$

with $D = \sum\limits_{p=1}^{N} [d(p)]^2$ and n_4 = number of 4-membered cycles in the graph. In the above formulae p has the same meaning as in Section 3.2.4.

The Δ-dependence of E_{pi} will be further elaborated in Section 3.6. and the D-dependence in Section 2.7.

3.4.   <u>TOPOLOGICAL FACTORS DETERMINING THE GROSS PART OF</u> E_{pi}

The fact that E_{pi} is proportional to the molecular size has been known for a long
time.[82,145] As will be shown in the subsequent chapter, the knowledge of the dependence of E_{pi} on the size of molecules is a prime practical importance and is closely
related to the problem of resonance energies.

The work of McClelland[119] was an important contribution towards the solution of
the problem. Namely, from the surprisingly simple two-parametric formula (3.31) it
follows that the gross part of E_{pi} is determined solely by molecular size. In other
words, about 95% of E_{pi} is determined by only two topological parameters--the number
of vertices N and the number of edges M.

The number of rings in a molecule is given by
$$R = M - N + 1. \qquad (3.73)$$
Therefore, one can write eq. (3.31) as

$$E_{pi} \approx a\sqrt{2N(N+R-1)}. \qquad (3.74)$$

This means that E_{pi} is proportional also to the number of rings present in the molecule.

For sufficiently large molecules, R-1<<N and E_{pi} should be almost linearly proportional to the number of vertices. Such linear dependences are observed in homo-
logous series (linear polyenes[62] annulenes[62] radialenes[76] polyacenes[42,88] etc.).

A practical consequence of eq. (3.74) is that it is not correct to compare the
E_{pi} values of molecules with same N but with different R values. For example, sometimes the stability of azulene (I in <u>Fig. 3.3.</u>) is compared with that of [10]-annulene
(II).

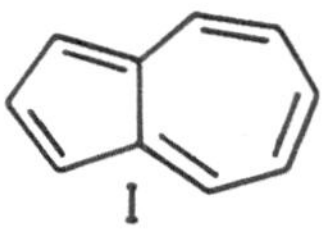

I
$E_{pi} = 13.36$

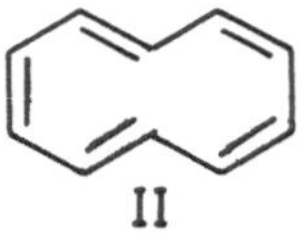

II
$E_{pi} = 12.94$

<u>Fig. 3.3.</u> E_{pi}'s of azulene (I) and [10]-annulene.

According to eq. (3.74) azulene should necessarily have greater E_{pi}; therefore any comparison of this kind is not reasonable.

We have seen that in the case that NBMOs are present in the molecule, $\tilde{N}$ is to be written instead of N in any formula for E_{pi}. The meaning of this is simply that the presence of NBMOs (i.e., zeros in the graph spectrum) decreases E_{pi}. The molecule can be formally treated as having $\tilde{N} = N - N_0$ conjugated centers. Since E_{pi} is rather strongly dependent on N, it is evident that the presence of NBMOs will substantially decrease E_{pi}. This is another argument in favour of extremely low stability of conjugated compounds, the molecular graphs of which have $N_0 > 0$.

After McClelland's work a whole class of further topological formulae (3.65), (3.69) and (3.70) confirmed the conclusion that <u>the gross part of E_{pi} is determined by M and N</u>. Numerical calculations show that M and N determine not less than 95% of E_{pi}. All other topological factors play, therefore, a seemingly marginal role.

However, it is to be noted that in all chemical applications we are interested not solely in energies but in <u>energy differences</u> (see the next chapter). The problem of the remaining 3-4% of E_{pi} is thus essential for chemistry.

Moreover, topological formulae containing only two parameters, M and N, are completely inapplicable for chemical discussions[63] One of their most striking failures is in making no difference between isomers.

3.5.   THE INFLUENCE OF CYCLES: THE HÜCKEL RULE

3.5.1.   General Considerations

The discovery of the fact that the chemical behaviour of cyclic conjugated compounds
is completely different, depending on whether the size of the cycles is 4m, 4m+1,
4m+2, or 4m+3 (m=integer), was one of greatest results which quantum theory gave to
chemistry. In this section we will be interested in different aspects of the problem
of how cycles influence E_{pi}.

The Hückel rule, which was originally discovered for annulenes[101] is well known.
It states that only annulenes with 4m+2 pi-electrons are stable. However, this rule,
at first glance, has nothing to do with E_{pi}, since it is based on the distribution
and degeneracy of MO energy levels.

On the other hand, it has also long been known both from quantum-chemical calculations and from experiments that Hückel rule-type regularities hold in conjugated
systems different than annulenes. One striking example is the fact that the most
stable conjugated compounds are the benzenoids, which contain only 6-membered rings.

Using perturbation theory, Dewar[32,33a] concluded that a regularity, completely
analogous to the original Hückel rule, holds for E_{pi}'s of arbitrary conjugated systems. Within the formalism of perturbational molecular orbital (PMO) theory this
extension of the Hückel rule was confirmed in Ref. 77.

We shall formulate the (extended) Hückel rule for E_{pi} in the following way:
every (4m+2)-membered cycle present in the molecular graph gives a positive and every
(4m)-membered cycle a negative contribution to E_{pi}.

However, it is not a priori clear what one should understand under the "contribution of a cycle" to E_{pi}. In the following text we shall explain how to extract the
contribution of a particular structural detail from the contributions of a large number of remaining, simultaneously acting, topological factors.

As a starting point we use eqs. (3.52) and (3.53). From the Sachs theorem,

$$U(x) = \sum_{j} \sum_{s} (-1)^{\frac{n(s)}{2} + c(s)} 2^{r(s)} x^{n(s)} \qquad (s \in S_{2j}) \qquad (3.75)$$

$$V(x) = \sum_{j} \sum_{s} (-1)^{\frac{n(s)-1}{2} + c(s)} 2^{r(s)} x^{n(s)-1} \qquad (s \in S_{2j+1}) \qquad (3.76)$$

where $n(s)$ is the number of vertices of the Sachs graph s, and the other symbols have their usual meaning. We see that every Sachs graph contributes either to the term U (if $s \epsilon S_{2j}$) or to the term V (if $s \epsilon S_{2j+1}$). Both terms U and V contribute further to E_{pi}. Therefore every Sachs graph can be understood as if contributing to the value of E_{pi}. This dependence is rather complicated, but analytically well defined. In the case of cycles a fortunate fact is that cycles are Sachs graphs. Hence the influence of cycles to E_{pi} can be followed straightforwardly.

Let us consider a cycle Γ of a graph G. Let $S_{j\Gamma}$ be the set of those Sachs graphs from $S_j = S_j(G)$ which contain the cycle Γ. Let $\overline{S}_{j\Gamma}$ be the complement of $S_{j\Gamma}$, that is the set of those Sachs graphs from S_j which do not contain the cycle Γ as a component. Of course,

$$S_{j\Gamma} \cup \overline{S}_{j\Gamma} = S_j \qquad (3.77)$$

$$S_{j\Gamma} \cap \overline{S}_{j\Gamma} = \phi. \qquad (3.78)$$

Then, we define by analogy to eqs. (3.75) and (3.76)

$$U_\Gamma(x) = \sum_j \sum_s (-1)^{\frac{n(s)}{2} + c(s)} 2^{r(s)} x^{n(s)} (s \epsilon \overline{S}_{2j,\Gamma}) \qquad (3.79)$$

$$V_\Gamma(x) = \sum_j \sum_s (-1)^{\frac{n(s)-1}{2} + c(s)} 2^{r(s)} x^{n(s)-1} (s \epsilon \overline{S}_{2j+1,\Gamma}) \qquad (3.80)$$

and consequently

$$E_{pi,\Gamma} = \tfrac{1}{2} \langle x^{-2} \ln(U_\Gamma^2 + x^2 V_\Gamma^2) \rangle. \qquad (3.81)$$

The difference $E_{pi} - E_{pi,\Gamma}$ is evidently the contribution of the cycle Γ to the value of E_{pi}.

The above algebra enables one to calculate the contribution of any cycle (and evidently, of any Sachs graph) to E_{pi}. In Ref. 64 the following result is given, enabling a much easier computation of $E_{pi,\Gamma}$.

Let analogy to Sachs theorem define

$$a_{j,\Gamma} = \sum_s (-1)^{c(s)} 2^{r(s)} (s \epsilon \overline{S}_{j,\Gamma}) \qquad (3.82)$$

and

$$P(G;\Gamma;x) = \sum_{j=0}^{N} a_{j,\Gamma} \, x^{N-j}. \tag{3.83}$$

Let the roots of $P(G;\Gamma;x)$ be $x_{1,\Gamma}, x_{2,\Gamma}, \ldots, x_{N,\Gamma}$. Let them be labeled in nonincreasing order, or so that

$$R_e(x_{1,\Gamma}) \geqslant R_e(x_{2,\Gamma}) \geqslant \ldots \geqslant R_e(x_{N,\Gamma}) \tag{3.84}$$

if the roots are complex. Then,

$$E_{pi,\Gamma} = \sum_{j=1}^{N} g_j \, R_e(x_{j,\Gamma}). \tag{3.85}$$

Formula (3.85) is completely analogous to eq. (3.1).

3.5.2. The Hückel Rule

In the present discussion of the Hückel rule we shall be interested only whether a cycle Γ has an increasing ($E_{pi} - E_{pi,\Gamma} > 0$) or decreasing ($E_{pi} - E_{pi,\Gamma} < 0$) effect. It will suffice to analyse the sign functions $(-1)^{n(s)/2+c(s)}$ and $(-1)^{(n(s)-1)/2+c(s)}$, since $2^{r(s)} x^{n(s)}$ for $n(s) =$ even and $2^{r(s)} x^{n(s)-1}$ for $n(s) =$ odd are always positive.

Let us further restrict ourselves to bipartite graphs. Then $S_{2j+1} = \phi$, $V(x)=0$ and $E_{pi} = \langle x^{-2} \ln U \rangle$.

We prove now the following theorem.[74a]

__Theorem__ (the Hückel rule). In a bipartite graph every $(4m+2)$-membered cycle has an increasing and every $(4m)$-membered cycle a decreasing effect on E_{pi}. In terms of the previously introduced notation, the theorem reads

$$E_{pi} - E_{pi,\Gamma} = \begin{cases} > 0 & \text{if the size of the cycle } \Gamma \\ & \text{is } 4m+2 \\[2mm] < 0 & \text{if the size of the cycle } \Gamma \\ & \text{is } 4m. \end{cases} \tag{3.86}$$

Let $n_j(s)$ be the number of j-membered cycles in a Sachs graph s. Define R_0, R_1, R_2 and R_3 as follows:

$$R_0(s) = n_4(s) + n_8(s) + n_{12}(s) + \ldots$$

$$R_1(s) = n_5(s) + n_9(s) + n_{13}(s) + \ldots$$

$$R_2(s) = n_6(s) + n_{10}(s) + n_{14}(s) + \ldots$$

$$R_3(s) = n_3(s) + n_7(s) + n_{11}(s) + \ldots . \qquad (3.87)$$

Then the following equality holds

$$(-1)^{\frac{n(s)}{2} + c(s)} = (-1)^{R_0(s)}. \qquad (3.88)$$

Remember that for bipartite graphs

$$n(s) = 2n_2(s) + 4n_4(s) + 6n_6(s) + 8n_8(s) + \ldots \qquad (3.89$$

$$c(s) = n_2(s) + n_4(s) + n_6(s) + n_8(s) + \ldots .$$

Then,

$$\frac{n(s)}{2} + c(s) = 2n_2(s) + 3n_4(s) + 4n_6(s) + 5n_8(s) + \ldots$$

$$= n_4(s) + n_8(s) + \ldots + 2(n_2(s) + n_4(s) +$$

$$+ 2n_6(s) + 2n_8(s) + \ldots). \qquad (3.90)$$

Therefore

$$(-1)^{\frac{n(s)}{2} + c(s)} = (-1)^{n_4(s) + n_8(s) + \ldots} = (-1)^{R_0(s)} . \qquad (3.91)$$

Proof. If the size of Γ is $4m+2$, we can distinguish two cases. Either a Sachs graph s from $S_{j,\Gamma}$ contains $(4j)$-membered cycles or not. If not, evidently $R_0(s)=0$, $(-1)^{n(s)/2+c(s)}=+1$ and therefore all those Sachs graphs have increasing contributions to E_{pi}.

If $s_1 \in S_{j,\Gamma}$ possesses at least one $(4j)$-membered cycle (say an 8-membered one), in the set $S_{j,\Gamma}$ there necessarily exist further two Sachs graphs s_2 and s_3 (see Fig. 3.4.).

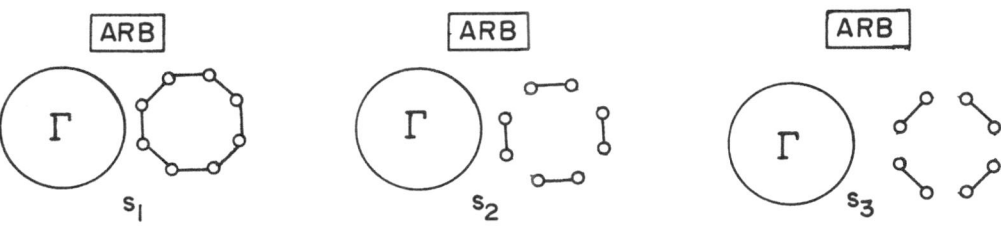

Fig. 3.4.

ARB denotes that the rests of s_1, s_2 and s_3 are arbitrary (e.g., they may contain also further (4j)-membered cycles).

Since

$$n(s_1) = n(s_2) = n(s_3)$$

$$c(s_1) = c(s_2) - 3 = c(s_3) - 3 \qquad (3.92)$$

$$r(s_1) = 4(s_2) + 1 = r(s_3) + 1$$

it turns out that the sum

$$\sum_{j=1}^{3} (-1)^{\frac{n(s_j)}{2} + c(s_j)} \, 2^{r(s_j)} \, x^{n(s_j)} \qquad (3.93)$$

is equal to zero. This means that the triplet of Sachs graphs s_1, s_2 and s_3 has in total a zero contribution to E_{pi}. Therefore, the total contribution of Sachs graphs from $S_{j,\Gamma}$ to E_{pi} is positive, that is to say that

$$E_{pi} - E_{pi,\Gamma} > 0 \text{ if the size of the cycle } \Gamma \text{ is } 4m+2. \qquad (3.94)$$

If the size of Γ is 4m, we can also distinguish two cases. Either a Sachs graph s from $S_{j,\Gamma}$ contains no (4j)-membered cycles except the cycle Γ, or contains further (4j)-cycles. If not, evidently $R_o(s)=1$, $(-1)^{n(s)/2+c(s)}=-1$ and therefore those Sachs graphs have a decreasing contribution to E_{pi}. A discussion analogous as before shows that all Sachs graphs s from $S_{j,\Gamma}$ such that $R_o(s) \geqslant 2$ can be matched into triplets, the total contribution of which is zero. As a final conclusion we have

$$E_{pi} - E_{pi,\Gamma} < 0 \text{ if the size of the cycle } \Gamma \text{ is } 4m. \qquad (3.95)$$

This completes the proof.

A detailed analysis[74] shows that the Hückel rule is valid also for nonalternant molecules.

3.5.3.   An Application: The Hückel Rule for Annulenes

According to the just-proved generalization of the Hückel rule, the pi-electrons should thermodynamically stabilize the [4m+2]- and destabilize the [4m]-annulenes. The reason that this simple fact was not noticed in the very early days of quantum chemistry is because E_{pi} is grossly determined by N and M (in annulenes, of course, N=M). Therefore E_{pi} of [N]-annulenes is a monotonously increasing function of N, which has nothing to do with the Hückel 4m+2 rule.

The Hückel rule-like behaviour of E_{pi}'s of annulenes is, however, easily exhibited if $E_{pi}-E_{pi,\Gamma}$'s are considered.

It is known[76] that $E_{pi}(N)$, the total pi-electron energy of the [N]-annulene is given by

$$E_{pi}(N) = \begin{cases} 4 \cot(\pi/N) \text{ is } N = 4m \\ \\ 4 \csc(\pi/N) \text{ if } N = 4m+2. \end{cases} \tag{3.96}$$

Using the facts that the [N]-annulene contains only one N-membered cycle (Γ), and therefore possesses exactly one cyclic Sachs graph, it can be deduced[60,64] that

$$E_{pi,\Gamma} = 2 \csc(\pi/2N). \tag{3.97}$$

Combing these two formulae, $E_{pi}(N)-E_{pi,\Gamma}(N)$ can easily be calculated. In Fig. 3.4. $E_{pi}(N)-E_{pi,\Gamma}(N)$ is presented as a function of N. The strong stability difference between [4m]- and [4m+2]-annulenes is evident. We call the reader's attention also to the fact that the difference between two classes of annulenes vanishes when N becomes large enough.

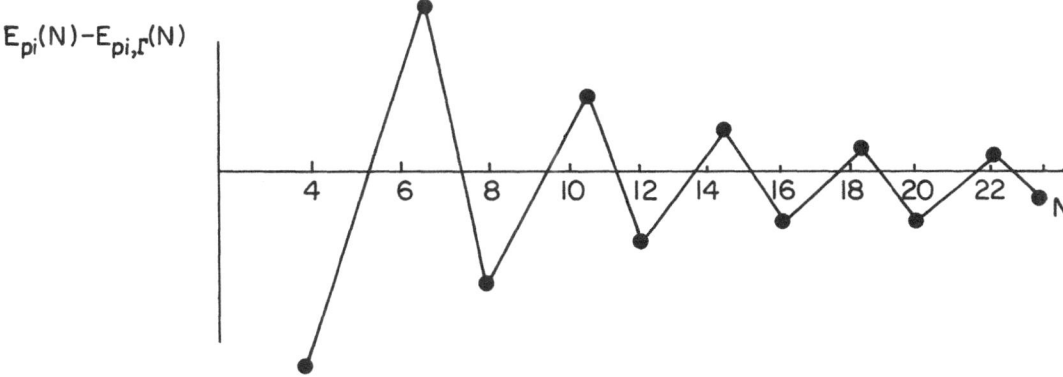

Fig. 3.4. The influence of the N-membered cycle on E_{pi} in [N]-annulenes.

3.5.4.   Extension of the Hückel Rule to Nonalternant Systems

Perturbation molecular orbital (PMO) arguments[32,77] indicate that odd-membered cycles should have small effect on E_{pi}. Graph-theoretical considerations, however, show that this is not exactly the case, although the effect of odd cycles is in general much smaller than the effect of even-membered ones.

The mathematical apparatus for the investigation of effects of odd cycles is completely analogous to that used for the proof of the Hückel rule. Therefore, in this section we shall outline only the most important details, while a complete analysis is to be found in Ref. 74.

We shall use eq. (3.52) as a starting point for the present discussion. In non-bipartite graphs both U and V are different from zero. We can focus our attention on the sign functions which fulfill the relations

$$(-1)^{\frac{n(s)}{2} + c(s)} = (-1)^{R_0(s)} (-1)^{\frac{R_1-R_3}{2}} \quad \text{for } n = \text{even} \qquad (3.96\,a)$$

$$(-1)^{\frac{n(s)-1}{2} + c(s)} = (-1)^{R_0(s)} (-1)^{\frac{\gamma-(R_1-R_3)}{2}} \quad \text{for } n = \text{odd} \qquad (3.97\,b)$$

where γ is the size of the smallest odd-membered cycle in the graph.

Because of the term $(-1)^{R_0}$, in both relations (3.96) and (3.97), it can be concluded[74] that the Hückel rule holds also for nonalternants. The effect of odd cycles is contained in the sign-term $(-1)^{(R_3-R_1)/2}$. Without going into further details we list here a set of rules which follow, more or less straightforwardly, from relations (3.96) and/or (3.97).

We shall call two odd cycles "of the same type" if the sizes for both are $4m+1$ or $4m+3$. A $(4m+1)$- and a $(4m+3)$-cycle will be said to be of the "opposite type". Note, however, that in graphs representing chemically relevant nonalternant topologies, 5- and 7-membered cycles occur almost exclusively.
Rule 1. If there is only one odd-membered cycle present in the molecular graph, there will be a small stabilization effect arising from its presence.

Numerical calculations confirm this conclusion but indicate, however, that the quantitative effect in conjugated systems containing one odd cycle (e.g. fulvene) is rather small.
Rule 2. Similarly, in the polycyclic case there exist also positive contributions to E_{pi} arising from all rings being of the same type as the smallest odd ring in the molecule. Odd rings of opposite type exhibit a destabilization effect.
Rule 3. A pair of either fused or disjoint odd-membered cycles causes stabilization if the cycles are of the opposite type and destabilization if they are of the same type.

A consequence of Rule 3 is a statement first formulated by Kruszewski and Krygowski[109]
Rule 4. A nonalternant molecule is relatively stable (aromatic) if the number of $(4m+1)$-membered rings is equal to the number of $(4m+3)$-membered rings.

3.6.   THE INFLUENCE OF KEKULÉ STRUCTURES

3.6.1.   Structure Count and Algebraic Structure Count

Kekulé structures of conjugated systems were introduced to theoretical chemistry much earlier and, of course, independently of HMO theory. Therefore, it was a considerable surprise to find that Kekulé structures and other closely related concepts play an important role in HMO theory. This was first noted by Longuet-Higgins[114] and later elaborated by various researchers. A survey of these relations is presented in Ref. 27 from a unified graph-theoretical viewpoint.

In the early stage of these investigations it seemed that certain quantities from HMO theory are related to the number of Kekulé structures (K, the structure count). However, it became clear[37] soon that this is the case only in benzenoid and acyclic hydrocarbons, while in the general case not K but another, more complicated topological function is required. This new quantity was first named "algebraic structure count" (ASC) by one[156,157] and later "corrected structure count" by other authors[90,91]. We shall use here the first of these two alternatives.

The problem of dependence of HMO quantities on K and ASC is not the topic of the present chapter. Therefore we shall give here few necessary results without proofs or detailed explanation. We refer the interested reader to Refs. 27, 91 and 157.

The notion of algebraic structure count can be introduced in the following way. For benzenoid hydrocarbons the equality

$$\det \underline{A} = (-1)^{N/2} K^2 \tag{3.98}$$

holds. When, however, nonbenzenoid structures are considered, eq. (3.98) is no more valid. We define ASC so that

$$\det \underline{A} = (-1)^{N/2} (ASC)^2 \tag{3.99}$$

for all conjugated systems. We see immediately that ASC = K for benzenoids. The definition (3.99) implies a number of interesting consequences.

First, in bipartite graphs a parity function Z_j can be introduced, such that $Z_j = +1$ or $Z_j = -1$ for the jth Kekulé structure and

$$\sum_{j=1}^{K} Z_j = ASC. \tag{3.100}$$

If $Z_j = +1$, the corresponding Kekulé structure is called even and if $Z_j = -1$, one speaks about <u>odd</u> Kekulé structures. Eq. (3.100) can be rewritten as

$$ASC = K^+ - K^-$$ (3.101)

with K^+ and K^- being the number of even and odd Kekulé structures, respectively. Note that $K^+ + K^- = K$.

There are several techniques available for the determination of the parity of Kekulé structures. We describe here the <u>superposition technique</u>.[27,69,72]

A superposition $k_i + k_j$ of two Kekulé graphs k_i and k_j is the graph possessing edges between all vertices which are adjacent at least in k_i or k_j. Of course, $k_i + k_j = k_j + k_i$ and $k_i + k_i = k_i$.

The three Kekulé graphs of benzcyclobutadiene graph and their superpositions are shown in <u>Fig. 3.5.</u>

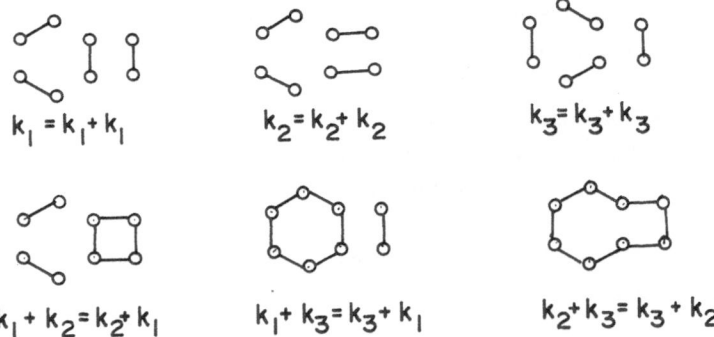

Fig. 3.5. Superpositions of Kekulé graphs for benzcyclobutadiene graph.

Let the Kekulé graphs of the graph G be k_1, k_2,..., k_K. It is easily seen that k_j's are in fact Sachs graphs. The same is true for their superpositions, and, moreover, if a graph is bipartite, all elements of S_N can be presented in the form $k_i + k_j$. Therefore, the only components of $k_i + k_j$ are cycles (of even size) and complete graphs of degree one.

<u>The superposition technique.</u> If the graph $k_i + k_j$ has an even number of cyclic components of the size $4m$, $Z_i = Z_j$. If the number of cyclic components of the size $4m$ in the graph $k_i + k_j$ is odd, k_i and k_j are of opposite parity, i.e., $Z_i = -Z_j$.

From the superposition technique we can determine the relative parity of Kekulé structures. The actual parity is then deduced from the fact that (by definition) $ASC \geqslant 0$.

As an example, consider benzcyclobutadiene. From <u>Fig. 3.5.</u> we see that $k_1 + k_2$ contains a 4-membered cycle, therefore $Z_2 = -Z_3$. (Consistently, $k_1 + k_2$ contains no $(4m)$-cycles and $Z_1 = Z_3$). From eq. (3.100) ASC $= Z_1 + Z_2 + Z_3 = Z_1 - Z_1 + Z_1 = Z_1$ and thus $Z_1 = 1$, $K^+ = 2$, $K^- = 1$ and finally ASC $= 1$.

It is seen from the superposition technique that <u>ASC reflects in an implicit way the ring structure of the molecule.</u> Since from the Hückel rule, $(4m)$-membered cycles decrease E_{pi}, it is to be expected that E_{pi} will be also a monotonously increasing function of ASC. In fact, we have proved this, because if $N_o = 0$, in formulae (3.69) and (3.70) $\Delta = (ASC)^2$.

3.6.2.   The Basic Postulate of Resonance Theory

One of the basic postulates of the resonance theory of conjugated molecules is that their   thermodynamic stability is an increasing monotonic function of the number of Kekulé structures that can be drawn for each compound. Largely because of the weight of accumulated empirical evidence the postulate continues to be accepted in spite of its known theoretical limitations.

The aim of the present section is to investigate the validity of the resonance theory postulate using a graph-theoretical approach. It appears that only a partial justification can be obtained and then only under restricted conditions. The present discussion will be limited to alternant hydrocarbons, but the conclusions can be readily transferred to heterocyclic analogues. The extension to nonalternants is also possible but implies addition difficulties and will not be considered here.

Using the identity

$$\det \underline{A} = \prod_{j=1}^{N} x_j \tag{3.102}$$

and the definition (3.99) of ASC, we obtain

$$(ASC)^2 = \prod_{j=1}^{N} \left| x_j \right| . \tag{3.103}$$

Assuming that no element of the graph spectrum is equal to zero, we obtain

$$2 \ln (ASC) = \sum_{j=1}^{N} \ln \left| x_j \right| . \tag{3.104}$$

In molecular graphs $|x_j| \epsilon (0,3)$ and we can approximate $\ln x$ in the interval $(0,3)$ by a <u>finite</u> polynomial $U_t(x) = \sum_{n=0}^{t} (-1)^{n+1} u_n x^n$. When this polynomial is substituted into eq. (3.104), one obtains simply

$$2 \ln (ASC) \approx \sum_{n=0}^{t} (-1)^{n+1} u_n W_n \tag{3.105}$$

where

$$W_n = \sum_{j=1}^{N} |x_j|^n . \tag{3.106}$$

Note that for even n, $W_n = L_n$ = number of loops of length n in the graph (see Section 3.2.3). Thus, the present analysis is closely related to the loop rule, Hence,

$$W_0 = N \tag{3.107}$$

$$W_2 = 2M. \tag{3.108}$$

From the fundamental identity it follows that

$$W_1 = E_{pi}. \tag{3.109}$$

Further it can be shown[75] that

$$W_3 \approx \frac{2E_{pi}}{N} (3M - \frac{E^2_{pi}}{N}). \tag{3.110}$$

For t = 3 eq. (3.105) gives

$$E_{pi} \approx \frac{c}{a} + \frac{b}{a} (\frac{c}{a})^3 \tag{3.111}$$

where

$$a = u_1 + 6u_3 M/N$$

$$b = 2u_3/N^2$$

$$c = 2 \ln (ASC) + u_0 N + 2u_2 M. \tag{3.112}$$

The coeffiecients u_0, u_1, u_2, u_3 have been determined by the least squares fitting method[75] and their optimal values are u_0 = 0.956, u_1 = 1.810, u_2 = 1.163, u_3 = 0.211. Numerical and other investigations along these lines were performed in Refs. 75, 158, 159. For the purpose of the present discussion it will suffice to establish that the approximate topological formula (3.111) (as well as formulae (3.69) and (3.70)) shows that E_{pi} is a monotonously increasing function of ln (ASC). (The actual analytical form of this function is highly dependent on the approximations used, and according to Ref. 59, cannot be established at all.)

The above discussion is slightly changed[159] in the case when det $\underline{A}$ = 0, that is when there are zeros in the graph spectrum. Then instead of eq. (3.103) we must consider the product π' of the absolute values of all nonzero graph eigenvalues

$$\Delta = \pi' \; x_j \;\; = \;\; (-1)^N \; (\prod_{j=1}^{N_+} \; x_j) \; (\prod_{j=N_++N_0+1}^{N} \; x_j). \tag{3.113}$$

$\ell n \Delta$ leads to terms W'_n of the form $W'_n = \Sigma' \, |x_j|^n$, where Σ' indicates the summation over nonzero elements only. Of course,

$$W'_0 = W_0 - N_0 = N - N_0 = \tilde{N}$$

$$\tag{3.114}$$

$$W'_n = W_n \quad \text{for } n = 1, 2, \ldots \; .$$

Thus we see that if $N_0 > 0$, in all topological formulae for E_{pi}, $(ASC)^2$ and N are to be substituted by Δ and $\tilde{N}$, while other topological parameters stay unchanged. The reader should compare this with the discussion in sections 3.2.4. and 3.3.2.

While ASC has a simple chemical interpretation, this is not the case for Δ. Nevertheless, the following formula holds[159] if $N_0 = 1$:

$$\Delta = \sum_{p=1}^{N} \; [ASC(G-p)]^2 \tag{3.115}$$

where $ASC(G-p)$ is the algebraic structure count of the graph $G-p$. Similarly, if $N_0 = 2$,

$$\Delta = \tfrac{1}{2} \sum_{p-1}^{N} \; \sum_{q=1}^{N} \; [ASC(G-p-q)]^2. \tag{3.116}$$

As a final conclusion of this section we point out that the predictions of the traditional resonance theory should not be generally valid because the parity of the Kekulé structures is not taken into account. The early apparent success of the resonance theory[154] rested on the fortunate fact that all Kekulé structures for benze= noid hydrocarbons and acyclic polyenes have the same parity[27,37]

The dependence of E_{pi} on ASC is logarithmic, so that ASC gives only a small, second-order contribution to E_{pi}. This is, of course, in agreement with the fact that E_{pi} is mainly determined by N and M. Thus, if one is interested in thermodynamic stability, the utility of the ASC is within a set of isomeric molecules.

3.7   THE INFLUENCE OF BRANCHING

As a final topological factor we discuss the molecular branching. Numerical experience shows that differences in branching cause only slight changes in E_{pi}. This effect usually cannot be noticed because of a number of other, much stronger effects which act simultaneously. Branching is therefore to be studied on conveniently chosen molecular graphs. Acyclic systems with Kekulé structure (K = 1) proved to be especially appropriate for this purpose.

There are numerous measures of branching proposed in the chemical literature.*
For discussing E_{pi} the following simple measure

$$D = \sum_{p=1}^{N} [d(p)]^2 \tag{3.117}$$

is sufficient. Let T, S and P be the number of vertices of degree three, two and one in a tree (= acyclic graph). Then

$$D = 9T + 4S + P \tag{3.118}$$

or after convenient transformations,

$$D = 2T + 4N - 6. \tag{3.119}$$

Hence D is (linearly) proportional to the number of branches in a tree.

The first indication that $\underline{E_{pi}}$ is a monotonously decreasing function of D came from an approximate formula of Hall.[82] Thereafter, this was confirmed by various other results[66,67] In particular, such a property of E_{pi} is reproduced by the formula (3.70). It has been established[66] that every branch decreases E_{pi} by nearly 0.1β. The fact proved in section 3.3.3 that the (nonbranched) path has maximal energy among trees is a nice illustration of the effect of branching.

3.7.1.   Violation of the Basic Postulate of Resonance Theory

In the previous section we proved that for benzenoid and acyclic systems a proportionality exists between E_{pi} and K. Hall found[83] that in certain cases the E_{pi}-K dependence is linear.

Thus, it is natural to ask whether the inequality K(G) > K(H) for two isomeric benzenoid systems G and H implied always $E_{pi}(G) > E_{pi}(H)$.

It could be proved[71] that this conjecture is not valid. The finding of an example which violates the conjecture is closely related with the difference in branching between G and H.

---

*For a survey of them see Ref. 129 and 133.

77

Suppose K(G) > K(H), but G is also more branched than H. It can occur that if the difference D(G) - D(H) is sufficiently large, the destabilization due to branching will predominate over the stabilization due to Kekulé structures.

Using such an argument, one has to compare[71] the E_{pi}'s of the graphs G_N and H_N (see Fig. 3.6).

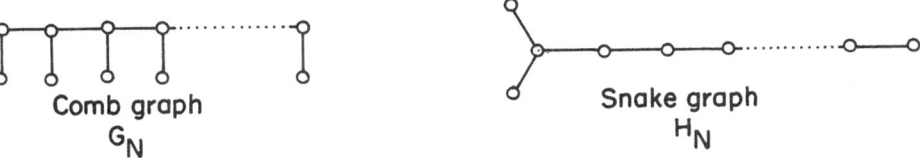

Fig. 3.6. Comb and snake graphs.

Really, $K(G_N) = 1$ and $K(H_N) = 0$, but G_N is much more branched than H_N,

$$D(G_N) = 5N - 10$$

$$D(H_N) = 4N - 4.$$

(3.120)

Since

$$E_{pi}(G_N) = \sum_{j=1}^{N} \sqrt{1 + \cos^2 \frac{2j\pi}{N+2}}$$

(3.120a)

$$E_{pi}(H_N) = 2 \cot g \frac{\pi}{2N-2}.$$

(3.121)

it is easy to find that the case $E_{pi}(H_N) > E_{pi}(G_N)$ will occur for $N \geqslant 18$. Hence, the molecules presented in Fig. 3.7. are examples of systems violating the basic postulate of resonance theory. Using the same method, many arbitrary isomeric structures with such a property can be designed.

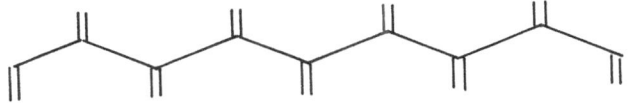

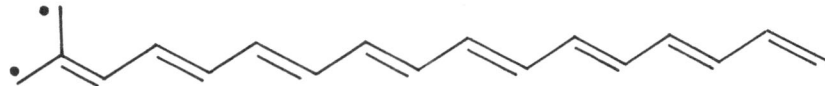

Fig. 3.7. The molecules violating the basic postulate of resonance theory

3.8   SUMMARY

In the present chapter we have analysed the topological factors which determine E_{pi}.
The following conclusions are the most interesting from a chemist's point of view.
A) The gross part (95% or more) of E_{pi} is determined just by the size of the molecule.
In the language of topology, E_{pi} is grossly determined by the number of vertices N
and edges M in molecular graph. Formula (3.31) reflects this clearly.
B) Since R = M - N + 1, E_{pi} is also proportional to the number of rings in the
molecule.
C) The above conclusions hold without modifications if there are no NBMOs in the
molecule. If, however, there are N_0 NBMOs (in the language of topology: N_0 zeros in
the graph spectrum), in all topological formulae one should set $\tilde{N} = N - N_0$ instead of
N. This means that the presence of NBMOs will substantially decrease E_{pi}.
D) The influence of even-membered cycles is summarized in the Hückel rule (section
3.5.2.). This rule is now proved in a rigorous mathematical fashion. The influence
of the odd-membered cycles to the value of E_{pi} is given in section 3.5.4. in a form
of a set of rules.
E) E_{pi} is a monotonously increasing function of the number of Kekulé structures only
for benzenoid and acyclic molecules provided $N_0 = 0$. Thus, the basic postulate of
resonance theory holds only within these restrictions.
F) For nonbenzenoid molecules the algebraic structure count, ASC, is shown to be proportional to E_{pi}. This proportionality is logarithmic. ASC is closely related to the
concept of parity of Kekulé structures. In the case $N_0 > 0$, E_{pi} is proportional to
another topological function Δ, the chemical interpretation of which is not simple.
G) Branching of the molecular skeleton slightly decreases E_{pi}, by nearly 0.1β for
every branch. Hence, in almost all chemical problems the effects of branching are
negligible.

4. RESONANCE ENERGY

4.1. INTRODUCTION

When total pi-electron energy is obtained as the result of a quantum chemical calculation, another problem remains unsolved. Namely, from the knowledge of E_{pi} one has to deduce chemically relevant predictions. This step in the utilizing of theoretical results is often tacitly assumed and its importance is sometimes forgotten.

In order to demonstrate to the reader the problem of interpretation, we give in <u>Fig. 4.1.</u> the E_{pi}'s of benzene (I)[25] and heptalene (II)[25]

$$E_{pi} = 8.00 \qquad\qquad E_{pi} = 15.62$$

<u>Fig. 4.1.</u> E_{pi}´s of benzene (I) and heptalene (II).

Without performing any further theoretical manipulation, we are unable to deduce anything from these numbers. In fact, the E_{pi} of heptalene was known long before its synthesis, but it was not evident whether (II) should be aromatic (as was often believed) or not (what is the experimental finding).

In the case of isomeric molecules the problem can be solved simply. Namely, the isomer with larger E_{pi} (in β units) should be (and is indeed) more stable, unless some severe steric hindrances are present.

Since E_{pi} is very much dependent on the size of the molecule, comparisons of E_{pi}'s of nonisomeric molecules are meaningless. However, the size of the molecule is usually not an essential factor in determining its chemical behaviour. Therefore, it is not only desirable but of the greatest practical importance to be able to compare calculated stabilities of unequal size molecules.

Resonance energy (or what is the same: delocalization energy) is a quantity introduced to solve the above problem. The conception of resonance energy can be expressed in the following "topological manner". We know that the gross part of E_{pi} is determined by molecular size (M,N), but we also know that this gross part of E_{pi}

is of minor importance for chemical predictions. Therefore, it would be desirable if one could separate the (M,N)-part of E_{pi} from the whole E_{pi}.

The general definition of resonance energy

$$RE = E_{pi} - E_{pi} \text{ (reference structure)} \tag{4.1}$$

can be now understood in a more abstract way, namely that E_{pi} (reference structure) is a topological function* which should contain all contributions to E_{pi} which can be ascribed to the presence of vertices and edges in the molecular graph.

---

*We will not consider any physical background behind this topological function. Therefore, we make no difference between "resonance" and "delocalization" energy.

4.2. CLASSICAL AND DEWAR RESONANCE ENERGIES

Since the reference structure in eq. (4.1) is a hypothetical and nonexisting entity, its choice is to a great extent arbitrary. Thus, resonance energy defined in the classical way, CRE, is given by

$$CRE = E_{pi} - 2n_=. \tag{4.2}$$

CRE is based on a reference structure which contains $n_=$ isolated double bonds with pi-electron energy of ethylene (2.0β). However, the CRE creterion is shown to fail in many cases because rather unstable molecules are predicted to be aromatic on the grounds of their CREs being large. For instance, the CRE of the aromatic benzene is 2.00β, while the CRE of the highly unstable heptalene is 3.62β.

This is easily understood knowing the McClelland formula (3.31). Since $n_= = N/2$,

$$CRE = E_{pi} - N. \tag{4.3}$$

From eq. (3.31) we have

$$CRE \approx (a\sqrt{2}-1)N \approx 0.3N \tag{4.4}$$

where we have also assumed that N>>R-1. This approximate relation shows that <u>CRE is grossly proportional to N</u>.

Several attempts have been made to redefine RE in order to obtain better agreement between theory and experiment. Dewar's suggestion[31,32,34] to use a reference structure that resembles an acyclic polyene and which has led to the introduction of the quantity called[2] "Dewar resonance energy" DRE,

$$DRE = E_{pi} - (n_= E_= + n_- E_-) \tag{4.5}$$

has proved to be the most valuable one. The DRE values obtained from eq. (4.5) correlated well with the experimental behaviour of conjugated systems.[34,36,92-95]

In eq. (4.5) $n_=$ and n_- are the number of double and single bonds in a Kekulé structure of the considered pi-system, whilst $E_=$ and E_- are parameters usually interpreted as the pi-energies of "polyene" double and single bonds, respectively.

The fundamental idea in the Dewar concept (which has been also expressed by Breslow and Mohacsi,[10] but not further elaborated) is to subtract from E_{pi} that energy part which corresponds to an <u>acyclic</u> polyene-like reference structure. Therefore, in the proper approach to this problem, one has to devise a parametrization scheme which should give zero (or nearly zero) DRE values for acyclic polyenes. Similarly, in such a parametrization DRE must contain all cyclic contributions to E_{pi}. DRE approaches of Dewar, and Hess and Schaad fulfill the above requirements only approximately. Furthermore, DREs can be evaluated (see formula (4.5)) only if the considered molecule possesses Kekulé structure. Therefore, radicals, ions, etc., are automatically excluded from DRE considerations.

4.3. TOPOLOGICAL RESONANCE ENERGY

4.3.1. The Mathematical Basis

Here we shall describe in some detail a variant of the DRE method that is free from the limitations of eq. (4.5), and which has some additional features.[65] Our variant is nonparametric and is directly related to the topology of the molecular pi-network. Thus, we wish to call it "topological resonance energy," TRE.

Using the results and formalism developed in the previous chapter, it is not difficult to offer a simple mathematical formulation of an "acyclic polyene-like reference structure". Evidently, "acyclic polyene-like" means that all structural details of a molecule except the presence of cycles should be taken into account.

Let S_j^{ac} be a subset of S_j, the elements of which have the property $r(s) = 0$. Hence, only those Sachs graphs are contained in S_j^{ac} which possess no cyclic components. Within the Sachs graph formalism, the ignoring of cycles means just the use of the set S_j^{ac} instead of S_j. Further, we define by analogy to the Sachs theorem,

$$a_j^{ac} = \sum_s (-1)^{c(s)} \, 2^{r(s)} \quad (s \epsilon S_j^{ac}). \tag{4.6}$$

The acyclic polynomial of a graph will be given as follows

$$p^{ac}(G;x) = \sum_{j=0}^{N} a_j^{ac} \, x^{N-j}. \tag{4.7}$$

Setting a_j^{ac} instead of a_j we obtain from eqs. (3.52) - (3.54)

$$E_{pi} \text{ (reference structure)} = \langle x^{-2} \ln \sum_j (-1)^j \, a_{2j}^{ac} \, x^{2j} \rangle \tag{4.8}$$

where we have used the fact that $S_{2j+1}^{ac} = \phi$ and therefore $a_{2j+1}^{ac} = 0$.

Furthermore, by analogy to eq. (3.48)

$$E_{pi} \text{ (reference structure)} = \langle x^{-2} \ln |H^{ac}(G;x)| \rangle \tag{4.9}$$

with

$$H^{ac}(G;x) = (ix)^N \, p^{ac}(G - i/x). \tag{4.10}$$

Then, we define the topological resonance energy TRE as

$$TRE = \langle x^{-2} \ln \left| \frac{H(G;x)}{H^{ac}(G;x)} \right| \rangle. \tag{4.11}$$

Another integral representation of TRE is obtained from eq. (3.58),

$$TRE = <\ln \left| \frac{P(G;ix)}{P^{ac}(G;ix)} \right| >. \tag{4.12}$$

Although both eqs. (4.11) and (4.12) are interesting topological relations, the actual computation of TRE is performed according to the relation

$$E_{pi} \text{ (reference structure)} = \sum_{j=1}^{N} g_j \, x_j^{ac} \tag{4.13}$$

which yields immediately

$$TRE = \sum_{j=1}^{N} g_j \, (x_j - x_j^{ac}). \tag{4.14}$$

In formulae (4.13) and (4.14) x_j^{ac} are the roots of $P^{ac}(G;x)$ labeled in a nonincreasing order. Relation (4.13) follows from eq. (3.85).

Thus, knowing the roots of $P(G,x)$ and $P^{ac}(G,x)$, one can calculate the TREs for arbitrary conjugated system G. Depending on the occupation numbers, one can calculate TREs for conjugated species with many arbitrary pi-electrons (any kind of radicals and ions). Ground and excited states could be equally treated by eq. (4.14).

It is now easy to show that if G is a tree, TRE(G) = 0. In other words, the following identity holds,

$$TRE \text{ (acyclic polyene)} = 0. \tag{4.15}$$

4.3.2.   The Computation of the Acyclic Polynomial

The evaluation of the acyclic polynomial $P^{ac}(G;x)$ is the crucial step in the computation of TRE. We describe and exemplify here a graphical procedure for the determination of P^{ac}.

According to the definition of a Sachs graph, S_j^{ac} contains only complete graphs of degree one, i.e., the $j/2$ edges which are nonincident in G. Therefore,

$$a_{2j}^{ac} = (-1)^j b_j$$

$$\tag{4.16}$$

$$a_{2j+1}^{ac} = 0$$

where b_j is the number of ways in which j nonincident edges can be selected in the graph $G(b_0 = 1)$. Therefore,

$$P^{ac}(G;x) = \sum_{j=0}^{N/2} (-1)^j b_j x^{N-2j}. \tag{4.17}$$

Let e be an edge of the graph G; G-e denotes the graph obtained by deletion of e from G. Similarly, G-(e) denotes the graph obtained by deletion of the both vertices incident to e. The following theorem about b_j's has been proved by Hosoya[97]

$$b_j(G) = b_j(G-e) + b_{j-1}(G - (e)). \tag{4.18}$$

Substitution of eq. (4.18) back into polynomial (4.17) leads to a recurrence relation for the acyclic polynomial,

$$P^{ac}(G;x) = P^{ac}(G-e;x) - P^{ac}(G-(e);x). \tag{4.19}$$

This equation is a generalization of the Heilbronner formula[87]

$$P(G;x) = P(G-e;x) - P(G-(e);x) \tag{4.20}$$

which is valid for acyclic polyenes only.

In order to apply the recurrence relation (4.19), remember that for acyclic graphs, $P^{ac}(G,x) \equiv P(G,x)$. Hence, by suitable choice of edges e in formula (4.19), P^{ac} can be expressed in terms of the characteristic polynomials of paths.

From eqs. (4.20) or (3.55) follows

$$P(P_N;x) = xP(P_{N-1};x) - P(P_{N-2};x) \tag{4.21}$$

which permits the easy calculation of these polynomials, starting with $P(P_1;x) = x$ and $P(P_2;x) = x^2 - 1$. For convenience, the characteristic polynomials of P_N's were tabulated in Ref. 87.

Let us calculate as an example the acyclic polynomial of naphtalene.

The acyclic polynomial of [10]-annulene is calculated easily as

$$p^{ac} \left(\text{} \right) = p^{ac} \left(\text{} \right) - p^{ac} \left(\text{} \right)$$

$$= P(P_{10}; x) - P(P_8 ; x)$$

Similarly,

$$p^{ac} \left(\text{} \right) = P \left(\text{} \right) = \left[P(P_4 ; x) \right]^2.$$

Finally,

$$p^{ac} \text{ (naphthalene)} = P(P_{10}, x) - P(P_8, x) - [P(P_4, x)]^2$$

$$= x^{10} - 9x^8 + 28x^6 - 35x^4 + 15x^2 - 1 -$$
$$(x^8 - 7x^6 + 15x^4 - 10x^2 + 1) - (x^4 - 3x^2 + 1)^2$$

$$= x^{10} - 11x^8 + 41x^6 - 61x^4 + 31x^2 - 3.$$

4.4.   TRE AS A CRITERION OF AROMATIC STABILITY.   CORRELATION WITH EXPERIMENTAL
        FINDINGS

DRE values of neutral conjugated molecules are abundantly available in the literature
due to efforts by Dewar and co-workers[34,36] (calculated in the framework of the SCF
pi-approximation) and by Hess and Schaad[92-95] (calculated within the HMO approximation). However, DRE cannot be used in a simple way for charged species. TRE, however, is not limited in its use. It could be applied equally well to neutral and
charged species as well as to excited states of conjugated structures. Therefore, one
should first investigate the reliability of TRE index by comparing it with DRE. The
comparison is made between TRE and DRE values of Hess and Schaad (simply because they
have also used Hückel theory in their work) for a large number of arbitrarily selected
neutral conjugated hydrocarbons and heterocycles. A least-squares fit for conjugated
hydrocarbons gives
$$TRE (PE) = 1.07 \, DRE (PE) - 0.009$$
with the correlation coefficient 0.97.
Similarly, for heteroconjugated molecules one obtains
$$TRE (PE) = 1.09 \, DRE (PE) - 0.005$$
with the correlation coefficient 0.97. TRE(PE) and DRE(PE) are topological resonance
energy per electron and Dewar resonance energy per electron, respectively. The above
analysis indicates that in most cases TRE predictions parallel those of DRE. Therefore, these two aromaticity indices, which are conceptually rather different, have
similar predictivity power.

The following threshold values of TRE are used for classifying conjugated compounds: those molecules with $TRE(PE) > 0.01\beta$ are considered aromatic, those -0.01
$< TRE(PE) < 0.01$ nonaromatic , and finally those compounds with $TRE(PE) < -0.01$ antiaromatic, reactive and unstable.

In order to illustrate the use of the TRE index we have selected 59 conjugated
structures which are reported in Table 4.1. Below each structure the TRE(PE) index
is given. In addition, for neutral compounds DRE(PE) indices are also reported. These
are given as the second number below each structure representing a neutral molecule.

Let us now discuss correlation between TRE predictions and experimental findings,
where available, for reported compounds in Table 4.1. Benzenoid hydrocarbons 1-4
have high values of TRE(PE) and are indeed aromatic compounds. DRE(PE) values of
these compounds are somewhat higher than corresponding TRE(PE), but lead, of course,
to the same prediction. Acenaphthylene (5) is also predicted aromatic by both indices
and this is supported by experiment.[40]

Predictions for nonbenzenoid alternant hydrocarbons 7-12 agree with their experimental behaviour. Thus, cyclobutadiene (7) is a highly unstable antiaromatic molecule which was only recently prepared in the gas matrix applying low temperature photolysis techniques[112,113] 3,4-dimethylenecyclobutene (8) is a very interesting compound. This compound is predicted antiaromatic (TRE(PE = -0.027β), but it is prepared and it represents one of a few cases of antiaromatic isolable compounds[102] Furthermore, 3,4-dimethylenecyclobutene has unusually high dipole moment (0.62D)[11] for an alternant structure. Benzocyclobutadiene (9) has a low TRE(PE) value (-01049β) and is a very unstable molecule indeed[16] Biphenylene (10) and its isomers 11 and 12 differ considerably in their stabilities. TRE(PE) index orders these three molecules according to their predicted aromatic stability as 10>12>11, biphenylene being the most stable and the linear isomer the least stable among them. This is well supported by experimental evidence[12]

Some of the nonalternant hydrocarbons 6, 13-24 are predicted aromatic (6, 16, 18, 21), others nonaromatic (13, 14, 22, 24) and antiaromatic (15, 17, 19, 20, 23). In the case of methylene-cyclopropene (6) the prediction is marginally supported by experimental data. Parent compound is not known, but its diphenyl derivative is prepared[5] and it is unstable unless it is carefully stored in the pure state. In addition, alkylmethylenecyclopropenes are remarkably stable at lower temperature, but polymerize at room temperature and in absence of solvent[144] Therefore, it appears that this compound is borderline aromatic indeed. 16 and 18 have both been known for some time and exhibit aromatic properties[56,107] Among the related molecules 15, 17 and 19 which are predicted antiaromatic, only 17 is prepared as an unstable molecule[!] Molecules 15-19 are bicyclic systems consisting of 4m+1 and 4m+3 rings. The topological rule[78] states that the bicyclic systems consisting of the odd-membered rings of the same size should be unstable while those consisting of the odd-membered rings of different sizes should be stable. Thus, molecules 15, 17 and 19 consist of two 3-, 5- and 7-membered rings, respectively, and are indeed unstable. Molecules 16 and 18 consist of 3- and 5-, and 5- and 7-membered rings, respectively, and are stable systems. Hence, this topological rule parallels predictions obtained by the TRE index.

Pentalene (20) and heptalene (22) are predicted antiaromatic, azulene (21) aromatic. These predictions agree with experiment; pentalene and heptalene are both unstable molecules[8,30] while azulene exhibits all characteristics of the truly aromatic species[105]

[18] annulene (23) is predicted by the TRE index nonaromatic. There are different predictions about its aromaticity available in the literature[35,79] some indicating [18]-annulene aromatic and others nonaromatic. Experimental evidence in this case is not conclusive [4,140-142]

Table 4.1. TRE(PE) values (in β units) of selected conjugated molecules. Second number underneath each molecule (if available) represents the DRE(PE) value of Hess and Schaad.

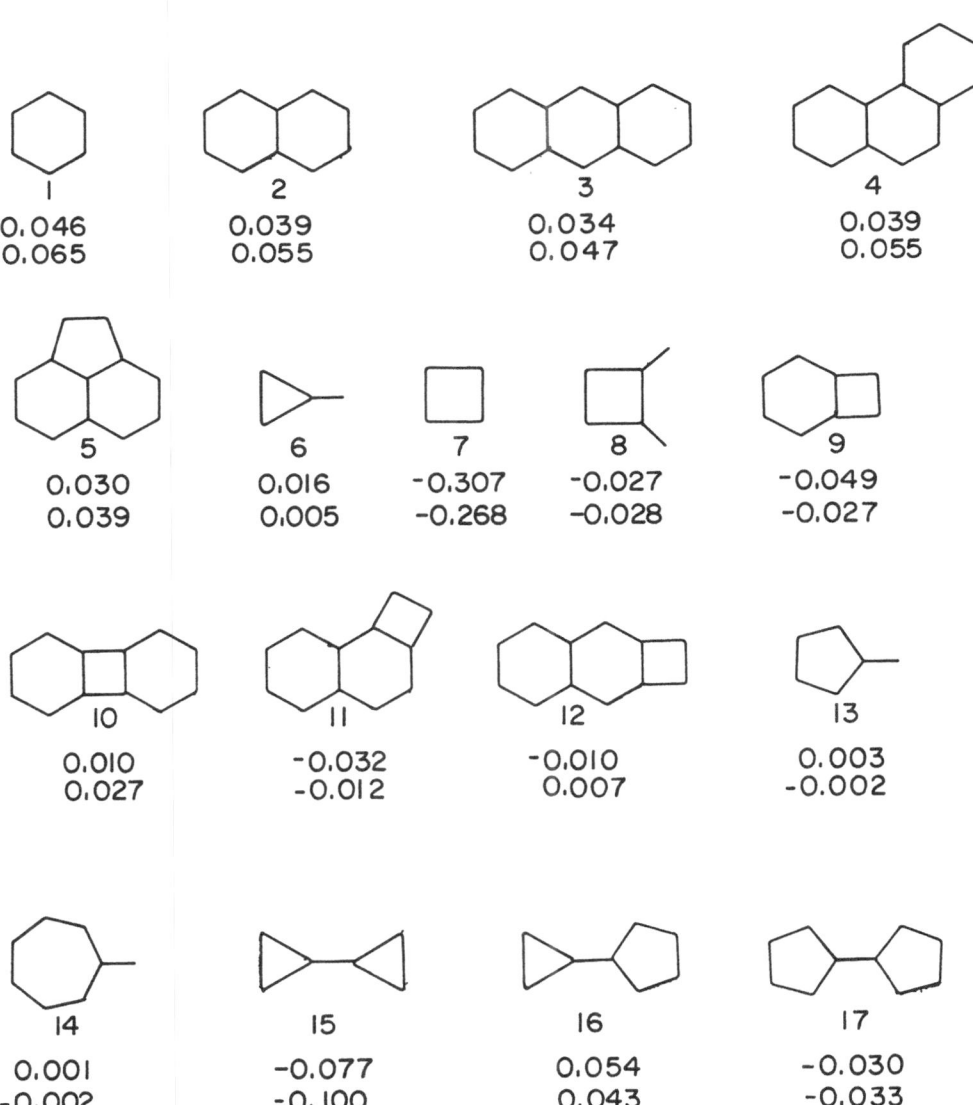

1	2	3	4
0.046	0.039	0.034	0.039
0.065	0.055	0.047	0.055

5	6	7	8	9
0.030	0.016	-0.307	-0.027	-0.049
0.039	0.005	-0.268	-0.028	-0.027

10	11	12	13
0.010	-0.032	-0.010	0.003
0.027	-0.012	0.007	-0.002

14	15	16	17
0.001	-0.077	0.054	-0.030
-0.002	-0.100	0.043	-0.033

Table 4.1. (cont´d)

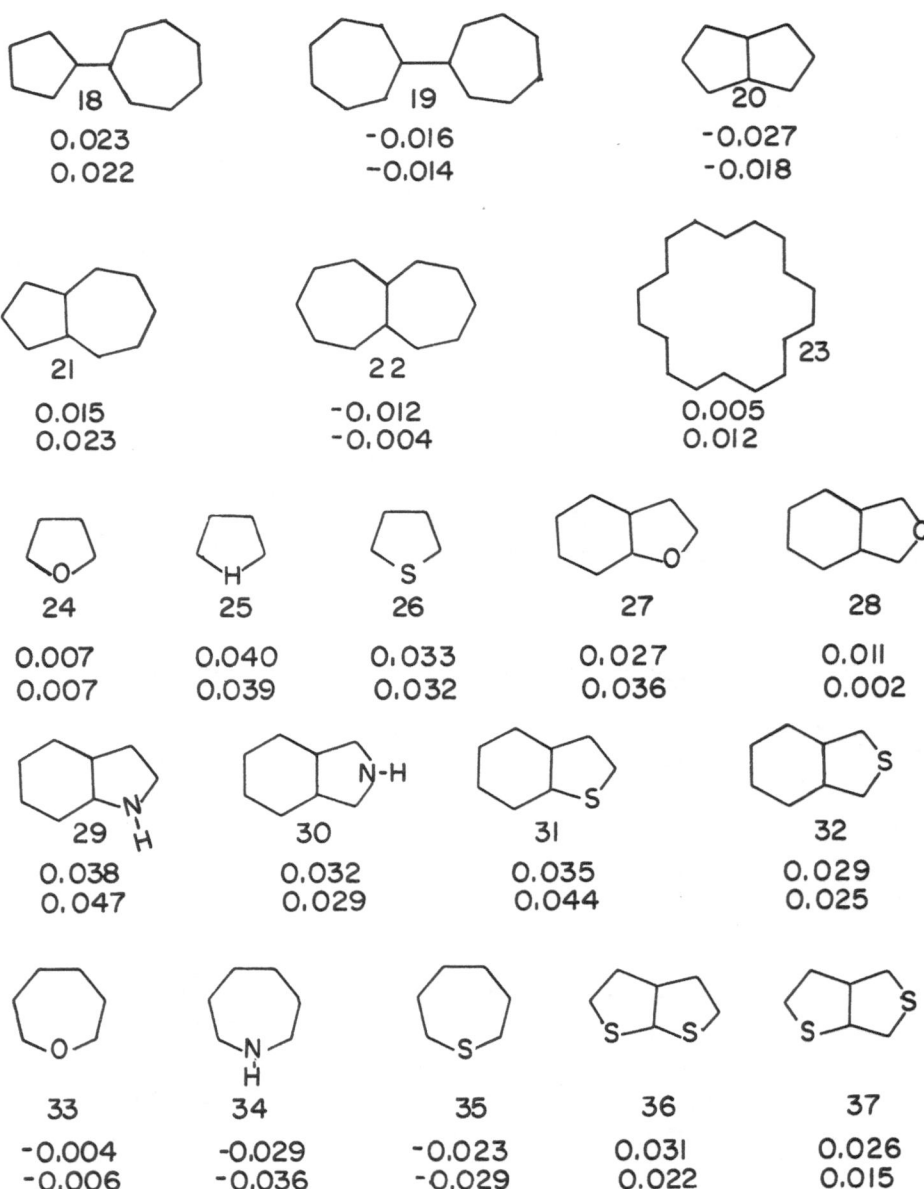

18
0.023
0.022

19
-0.016
-0.014

20
-0.027
-0.018

21
0.015
0.023

22
-0.012
-0.004

23
0.005
0.012

24
0.007
0.007

25
0.040
0.039

26
0.033
0.032

27
0.027
0.036

28
0.011
0.002

29
0.038
0.047

30
0.032
0.029

31
0.035
0.044

32
0.029
0.025

33
-0.004
-0.006

34
-0.029
-0.036

35
-0.023
-0.029

36
0.031
0.022

37
0.026
0.015

91

Table 4.1. (cont´d)

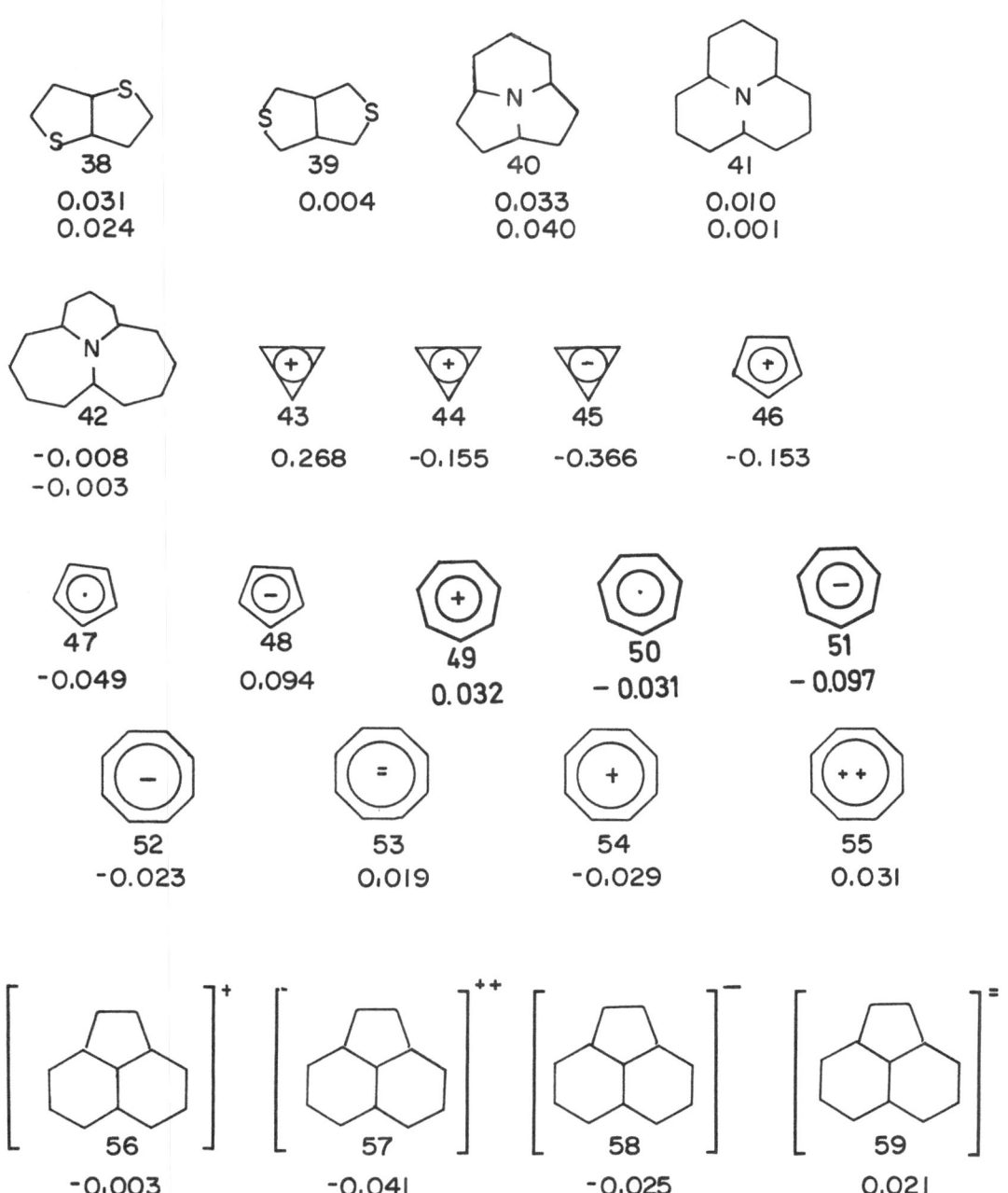

38 0.031 0.024	**39** 0.004	**40** 0.033 0.040	**41** 0.010 0.001
42 -0.008 -0.003	**43** 0.268	**44** -0.155	**45** -0.366	**46** -0.153
47 -0.049	**48** 0.094	**49** 0.032	**50** -0.031	**51** -0.097
52 -0.023	**53** 0.019	**54** -0.029	**55** 0.031
56 -0.003	**57** -0.041	**58** -0.025	**59** 0.021

In the case of molecules 24-26 the TRE index correctly differentiates between furan, 24, (nonaromatic), and pyrrole, 25, and thiophene, 26, (both aromatic). Similarly, the TRE index predicts benzofuran (27), benzopyrrole (29), and benzothiophene (31) more stable than corresponding isomeric molecules, isobenzofuran (28), isobenzopyrrole (30) and isobenzothiophene (32). The abundant experimental evidence supports this theoretical result. In addition, oxepine (33) is predicted nonaromatic, axepine (34) and thiepine (35) antiaromatic. This is consistent with experimental findings; the oxygen-containing system (oxepine) is a somewhat stabler species than either N-containing (azepine) or S-containing 7-membered ring (thiepine)[80,96,116]

Thiophthenes (36-39) are arranged by the TRE index in the following stability order: (36) $\approx$ (38) > (37) >> (39). Experimental fact is that isomers 36-38 are synthesized[55,162] and they differ somewhat in their stabilities, 1,4- and 1,6-isomers being more stable than the 1,5-isomer. 2,6-thiophthene (39) is, however, observed as a transient species and a very reactive intermediate[13] Here we should emphasize that in TRE calculations for heterocyclics the parameter scheme of Hess and Schaad[95] is adopted. However, these parameters were not fitted for nonclassical structures as, for example, is 2,5-thiophthene, and thus parameters used may not be convenient for such systems. This is probably the reason why the TRE value of 2,5-thiophthene is nonaromatic instead of expected antiaromatic.

Cyclazines 40-42 are predicted to be of different stability, relative stability order being 40 > 41 > 42. Cycl [3.2.2] azine (40) is predicted aromatic, two other compounds nonaromatic, cycl [4.4.3] azine (42) being the less stable of the two. Experimental evidence is in accord with these predictions. Cycl [3.2.2] azine is found a stable and aromatic molecule[160] Cycl [3.3.3] azine (41) has only recently been made, but it is a rather reactive compound[45] There is no evidence in the literature about the preparation of cycl [4.4.3] azine or its derivatives.

The TRE concept is very useful for studying aromaticity of conjugated redicals and ions. Chemistry of conjugated radicals and ions has developed rapidly in the last decade due to the improvements of the preparative techniques[103] The TRE index applied to nomocyclic ring systems C_m (m=3,5,7,8) leads to predictions consistent with those obtained from the Hückel rule[101] Thus, cyclopropenyl cation, $C_3H_3^+$ (43), cyclopentadienyl anion, $C_5H_5^-$ (48), tropylium cation, $C_7H_7^+$ (49), cyclooctatetraenyl ication, $C_8H_8^{++}$ (55), and dianion, $C_8H_8^=$ (53), representing 4m+2 systems, are predicted to be stable aromatic ions. All these ions are stable in solution[49,50,123] TRE predicts the following stability order among C_3, C_5, C_7 and C_8 ring species: $C_3H_3^+ > C_3H_3^{\cdot} > C_3H_3^-$, $C_5H_5^- > C_5H_5^{\cdot} > C_5H_5^+$, $C_7H_7^+ > C_7H_7^{\cdot} > C_8H_8^{\cdot}$, and $C_8H_8^{++} \approx C_8H_8^= > C_8H_8^+ \approx C_8H_8^-$. These results are fully in agreement with experimental observations[49,123] Cyclooctatetraene (COT) ions have been especially thoroughly studied:

(COT)$^-$, (COT)$^=$ were prepared[104] in 1960, (COT)$^{++}$ only recently[123] while (COT)$^+$ is still unknown. However, the difficulties in the preparation of (COT)$^+$ should be expected since its predicted stability is comparable to that of (COT)$^-$ which is only observed as a short-lived species in solution.

We report also results for acenaphthylenyl cations and anions 56-59. The TRE index produces the following stability order of these species: parent compound (5) > (59) > (56) > (58) > (57). Aromatic species are predicted parent compound (5) and acenaphthylenyl dianion (59), nonaromatic only cation (56), acenaphthylenyl dication (57) and anion (58) antiaromatic structures. Molecules 5 and 59 are both prepared[40,107a] while other acenaphthylenyl ions have not been observed yet.

4.5.   CONCLUDING REMARKS

The TRE concept is the result of the translation of Dewar's definition of the reference structure into the language of graph theory. Analysis shows that the TRE index parallels DRE in predicting the aromatic stability of a large number of neutral conjugated systems. In addition, TRE is used for studying conjugated radicals and ions without further adjustment of the theory. Furthermore, we point out that in all studied cases for which there were experimental data available, agreement between the theory and experimental findings was achieved. This leads to the conclusion that the TRE index may be used as a powerful theoretical index for predicting aromatic properties of arbitrary conjugated systems.

5.   REACTIVITY OF CONJUGATED STRUCTURES

5.1.   LOCALIZATION ENERGY

The reactivity of conjugated structures may be considered at various levels. The
most fundamental level is a <u>dynamical</u> study of the elementary reaction process, that
is to say, the evaluation of reaction cross sections. However, the present state of
development allows us to consider chemical dynamics of only the simplest reactions.
In addition, for chemical dynamics reliable potential surfaces are necessary, but
calculating them also presents a formidable problem. However, the dynamic approach
to chemical reactivity represents the level of the future; the present level and the
level of the past is the well-known <u>reactivity indices method</u>. This method was
shown to be very useful for predicting the relative reactivities of a series of
structurally closely related molecules. The reactivity indices method appeals directly to chemical intuition and it is formulated in terms of elementary concepts
of quantum mechanics.

Here we wish to discuss the reactivity index called <u>localization energy</u>.[146] This
index is based in the model introduced by Wheland[153] and may be used for predicting
the aromatic substitution reactions. Wheland has postulated that the transition
state in aromatic substitution reactions resembles a sigma-complex. A sigma-complex
is schematically presented in <u>Fig. 5.1.</u>

Fig. 5.1. Sigma-complex.

In the sigma-complex atom r in the substratum is at the same time connected to a de-
parting hydrogen atom and attacking group X. Thus, it has approximately a tetrahe-
dral configuration and is no longer contributing to a conjugation within the aromatic
ring. Hence, the conjugated part of the substratum has one atom less. We can also say
that during the substitution reaction, an electron on atom r is localized. The energy necessary
for this localization (ΔE_{pi}) may be calculated using the following formula:

$$\Delta E_{pi} = E_{pi} \text{ (molecule)} - E_{pi} \text{ (S)}. \tag{5.1}$$

In the literature this localization energy is usually denoted by the symbol L_r. It actually represents the change in the pi-electron energy of a conjugated molecule in the course of the forming sigma-complex. The lower this energy the lower will be the energy of the transition state, and the aromatic substitution reaction will proceed with lesser energy expense. Therefore, the lower the L_r-value, the lower (energy) the barrier for the substitution at the corresponding position r.

E_{pi} (molecule) and $E_{pi}(S)$ may be calculated using various MO methods presently available. However, for this kind of theoretical approach to chemical reactivity the Hückel theory is a very convenient one, because it is simple to use. This is a very attractive feature of any theory if it pretends to appeal to experimental chemists.

Let us evaluate, as an example, localization energies for two positions (α and β) of biphenylene (1) (see Fig. 5.2.).

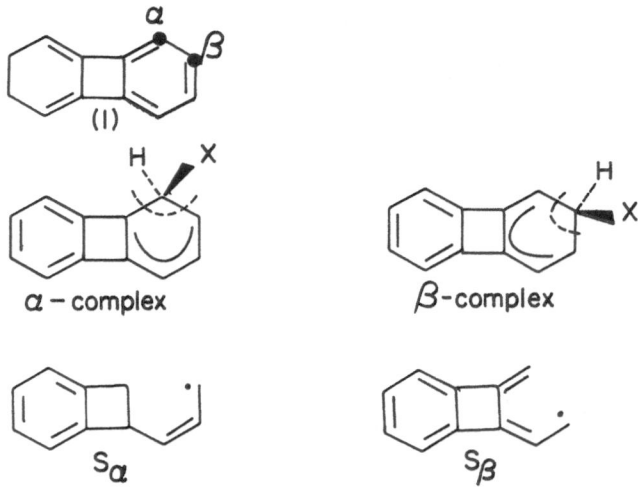

Fig. 5.2. Biphenylene sigma-complexes generated in the course of the radical attack.

First, we construct sigma-complexes for two positions of biphenylene available for the attack of a substituent. The substituent could be nucleophile, electrophile, or radical. This means that in the first case no pi-electrons will be necessary to be localized at atom r, in the second case two, and in the last only one pi-electron, respectively. Expressions for calculation of L_r are as follows:

$$L_r^N = E_{pi} \text{ (biphenylene)} - E_{pi} (S^-) \qquad (5.2)$$

$$L_r^E = E_{pi} \text{ (biphenylene)} - E_{pi} (S^+) \qquad (5.3)$$

$$L_r^R = E_{pi} \text{ (biphenylene)} - E_{pi} (S^\cdot). \qquad (5.4)$$

In the second step, HMO calculations for biphenylene and S-fragments are carried out.

$$E_{pi} \text{ (biphenylene)} = 16.71, \ E_{pi} (S_\alpha^-) = E_{pi} (S_\alpha^+) = E_{pi} (S_\alpha^\cdot) = 14.10,$$

$$E_{pi} (S_\beta^-) = E_{pi} (S_\beta^+) = E_{pi} (S_\beta^\cdot) = 14.15.$$

These numbers are in β units, while α is taken as the zero-energy point.

$$L_\alpha^N = 16.51 = 14.10 = 2.41$$

$$L_\beta^N = 16.51 - 14.15 = 2.36.$$

Here we learn that for alternant conjugated structures always $L_r^N = L_r^E = L_r^R$ because the S-fragment derived from the alternant hydrocarbon contains one NBMO. Predictions reached by the localization energy index--$L_\alpha < L_\beta$; $\alpha > \beta$--are in full agreement with experimental findings[12]

The L-index has been a very useful one for predicting the most probable position on a particular conjugated structure for nucleophilic, electrophilic or radical attacks[32,146] However, it has one practical drawback from the point of a potential user: the experimental chemist. While HMO[25] or SCF MO[39] energies and orbitals are tabulated for a large number of conjugated molecules, it is necessary to carry out either Hückel or SCF MO calculations for every fragment derived from the sigma-complex considered. For example, in the case of azulene there are five such fragments possible, and in the case of 1,2-benzpyrene even 12 different fragments. This is not very attractive for the user. He needs quicker methods for the evaluation of indices, even if such a method is approximate. Dewar[32] has, in fact, developed an approximate method for computing localization energies of alternant structures based on the application of the perturbation theory to HMOs[33a]

5.2.   DEWAR NUMBER

Dewar has shown how for alternant hydrocarbons one can easily obtain approximate
localization energies from the coefficients of a non-bonding molecular orbital (NBMO).
Approximate localization energy is denoted by N_S and is called the <u>Dewar number</u>.

The Dewar number may be derived in the following way. If two united conjugated
fragments R and S produce conjugated hydrocarbon RS, the Hückel energy of a molecule
RS is approximately given by,

$$E_{RS} \approx E_R + E_S + 2(\sum_{r-s} C_{or}^R \, C_{os}^S) \qquad\qquad (5.5)$$

where E_R and E_S are HMO energies of fragments R and S, respectively, and C_{or}^R and
C_{os}^S coefficients at atomic orbitals $\phi_r(R)$ and $\phi_s(S)$, respectively, in NBMOs of fragments R and S. Summation is over all r-s bonds. Let us illustrate the above by an
example. Consider two allyl fragments which by connecting terminal atoms give benzene (see <u>Fig. 5.3.</u>).

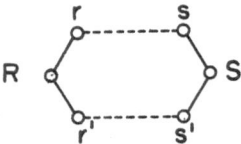

<u>Fig. 5.3.</u> PMO creation of benzene.

The Hückel energy of allyl fragment is 2.82 (remember Hückel energy values are reported in β units, while α is taken as the zero-energy point), and the NBMO of allyl
radical is given by

$$\psi_{NBMO}(\text{allyl}) = \frac{1}{\sqrt{2}} \phi_r - \frac{1}{\sqrt{2}} \phi_{r'} . \qquad\qquad (5.6)$$

Employing relation (5.5) the following approximate HMO energy of benzene is obtained:

$$E_{RS}(\text{benzene}) \approx E_R(\text{allyl}) + E_S(\text{allyl}) + 2(C_{or}^R \, C_{os}^S + C_{or'}^R \, C_{os'}^S)$$

$$2.282 + 2[(1/\sqrt{2})(1/\sqrt{2}) + (-1/\sqrt{2})(-1/\sqrt{2})] = 7.64.$$

This number compares with 8 for full calculation on benzene. The reason why the approximate value is lower than the true value lies in the fact that Dewar's procedure
neglects all interactions between fragments R and S except those among two NBMOs.

However, Dewar has shown that there is a linear relationship between E_{RS} values and Hückel energies.

In a very special case of substitution reactions, one fragment may be conjugated as part of the sigma-complex and other just the atom singled out by the substituent attack. This is illustrated below.

For this case the relation (5.5) reduces to a simpler form (note that $C_{os} = 1$)

$$E_{RS} \approx E_R + E_S + 2(C_{or}^R C_{os}^S + C_{or'}^R C_{os'}^S) \tag{5.7}$$

$$E_{RS} - (E_R + E_S) \approx 2(C_{or}^R + C_{or'}^R) \tag{5.8}$$

thus showing that the localization energy is approximately given by the quantity

$$N_S = 2(C_{or}^R + C_{or'}^R) \tag{5.9}$$

which is called the Dewar number. Eq. (5.9) is valid for C_{oj}s being the coefficients normalized NBMOs.

Let us illustrate the calculation of the Dewar number for anthracene. An attack on anthracene can proceed through the three sigma-complexes shown below.

Now, the first step is to construct NBMOs corresponding to conjugated parts of sigma-complexes. This can be simply done by following the sum rule,[114] i.e., the sum of coefficients around any atom in the NBMO must be zero.

$$\sum_{p \neq q} C_p = 0; \quad q = 1,2,\ldots,N. \tag{5.10}$$

The summation is over all atoms p joined to the atom q.

The use of eq. (5.10) is illustrated for a peri-complex. Unnormalized NBMO is considered.

Normalization of the NBMO leads to an explicit value of a.

$$a^2 + a^2 + a^2 + 4a^2 + a^2 + a^2 + a^2 = 1, \quad a = 1/\sqrt{10}.$$

Thus, the NBMO for a peri-complex is given by,

Similarly, the NBMOs for an α- and β-complex are given below.

The next step is the computation of Dewar numbers by means of eq. (5.9).

$$N(\text{peri}) = 2\left(\frac{1}{\sqrt{10}} + \frac{1}{\sqrt{10}}\right) = 1.26$$

$$N(\alpha) = 2\left(\frac{1}{\sqrt{26}} + \frac{3}{\sqrt{26}}\right) = 1.57$$

$$N(\beta) = 2\left(\frac{1}{\sqrt{18}} + \frac{3}{\sqrt{18}}\right) = 1.89.$$

The following reactivity prediction is thus obtained:

$$N(peri) < N(\alpha) < N(\beta)$$
$$peri- > \alpha- > \beta-.$$

Since the N is the smallest for the peri-position, the substitution at this position should be most favoured. This is in accord with experimental evidence.

The weakness of Dewar's approach is that it cannot be used for nonalternant hydrocarbons because their fragments do not in general possess NBMOs.

5.3.   TOPOLOGICAL APPROACH TO LOCALIZATION ENERGY

In Section 5.1. it was shown how localization energies can be obtained from the direct calculations. But in the computerized approach we do not learn much about the connection between structural features of the molecule and molecular reactivity. One way to analyze such a relationship is to apply graph-theoretical concepts to aromatic substitution and localization energies.

In the language of graph theory the sigma-complex, generated in the course of substitution reactions, is represented by a graph obtained after deleting the appropriate vertex and incident edges from the molecular graph. For example, biphenylene (1) and its α-(2) and β-complexes (3) are represented by graphs 7-9, respectively.

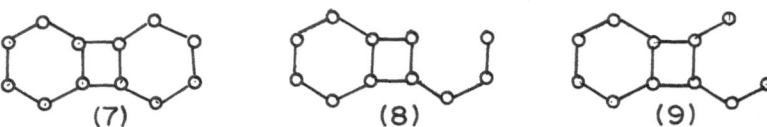

Graphs 8 and 9 have one vertex and two edges less than graph 7 corresponding to the parent molecule. Graphs 8 and 9 are bipartite (bicolourable) graphs and since they have an odd number of vertices, their graph spectra must contain at least one zero element, $N_O = 0$ (N_O denotes the number of zeros in the spectrum of the molecular graph). However, in our case these graphs contain exactly one zero element in their spectra, i.e., $N_O(8) = N_O(9) = 1$. Graph 7 has no zero element in its spectrum, i.e., $N_O(7) = 0$.

Expression (5.1) for the evaluation of localization energy can be rewritten using graph-theoretical language.

$$L_r = E_{pi}(G) - E_{pi}(G_r) \qquad\qquad (5.11)$$

where G is the molecular graph and G_r = G-r is the subgraph obtained by deletion of the vertex r and all the edges incident to it.

Earlier in section 3. the topological formula for total pi-electron energy was derived.

$$E_{pi} = A(N-N_O) + B \ln (acs) + C M \qquad\qquad (5.12)$$

where N and M represent the number of vertices (atoms) and edges (bonds); asc is the algebraic structure count of molecule containing NBMOs and is defined as follows:

$$(asc)^2 = \prod_{j=1}^{N}{}' |x_j| \qquad (5.13)$$

where $\prod'$ means multiplication over all nonzero elements $(x_j, j=1,2,...)$ of the graph spectrum. (asc) actually represents the generalization of the algebraic structure count (ASC) concept

$$(ASC)^2 = \prod_{j=1}^{N} |x_j|. \qquad (5.14)$$

It is evident that in the case of $N_o = 0$,

$$ASC = asc. \qquad (5.15)$$

A(0.913), B(0.765) and C(0.347) are constants derived by a least-squares fitting of eq. (5.13) to an arbitrarily selected group of conjugated hydrocarbons and ions.

$E_{pi}(G)$ and $E_{pi}(G_r)$ can be explicitly written by means of eq. (5.12),

$$E_{pi}(G) = A N + B[\ln ASC(G)] + C M \qquad (5.16)$$

$$E_{pi}(G_r) = A[(N-1)-1] + B \ln[asc(G_r)] + C(M-2) \qquad (5.17)$$

where it is assumed that $N_o(G) = 0$ and $N_o(G_r) = 1$.

Substituting eqs. (5.16) and (5.17) back into eq. (5.11) the following expression for localization energy is obtained:

$$L_r = 2(A + C) + B \ln[ASC(G)/acs(G_r)]. \qquad (5.18)$$

Since A, B and C are constants, the localization energy for the position r is determined by a unique topological paramenter,

$$D_r = \frac{ASC(G)}{asc(G_r)} . \qquad (5.19)$$

The relative reactivity of two positions r and s (not necessarily on the same substratum) is determined by the difference in the corresponding localization energies,

$$\Delta L_{rs} = L_r - L_s. \qquad (5.20)$$

The substitution of eq. (5.18) into (5.20) leads to the expression

$$\Delta L_{rs} = B \ln(D_r/D_s) \qquad (5.21)$$

Let us use the assumption of Dewar[38] that L_{rs} can be identified with the difference in the free energy of activation,

$$k_r/k_s \approx \exp(-\Delta L_{rs}/RT) \tag{5.22}$$

where k_r and k_s are the rate constants for an aromatic substitution reaction on the atoms r and s, respectively, T(K) is the absolute temperature, and R(8.31431 JK^{-1} mol^{-1}) the ideal gas constant.

By substituting eq. (5.21) into (5.22) we obtain finally,

$$k_r/k_s = (D_s/D_r)^\varsigma \tag{5.23}$$

where ς = B/RT.

From this equation we learn that the position with larger (asc) value will be more reactive,

$$D_s/D_r = \text{asc}(G_r)/\text{asc}(G_s) \tag{5.24}$$

(asc) is a quantity closely analogous to the number of resonance forms of the corresponding sigma-complex, i.e., for N_o=1

$$(\text{asc})^2 = \sum_{r=1}^{N} [(ASC)_r]^2 \tag{5.25}$$

where $(ASC)_r$ is the algebraic structure count of the graph G-r.

If all resonance forms are of the same parity, the more reactive position is one which has a larger number of resonance structures (K) in the corresponding sigma-complex. However, in the general case the parity of the resonance forms should be considered. Therefore, the topological rule for predicting substituent orientation could be summarized as follows: <u>the more reactive position towards aromatic substitution is that one which has a larger algebraic count of resonance structures in the corresponding sigma-complex.</u>

Let us apply this rule to biphenylene α- and β-complexes. Corresponding K and (asc) values are as follows: K(α)=11, K(β)=10, $[\text{asc}(\alpha)]^2$=9, and $[\text{asc}(\beta)]^2$=12. The traditional form of resonance theory clearly gives the wrong prediction by favouring α-position over β-position since experiment[12] shows that substituents attack more readily the β-position. However, the prediction based on the topological rule, which favours the β-position, is thus completely verified with experimental evidence.

5.4   TOPOLOGICAL ASPECT OF DEWAR NUMBER

For <u>unnormalized</u> NBMOs the Dewar number can be rewritten in the form

$$N_S = 2(C_{or} + C_{or'})/ \sum_{j=1}^{N-1} C_{oj}^2. \tag{5.26}$$

Coefficients C_{oj} of NBMOs can be obtained using Dewar´s approach[32] (being integer numbers). For anthracene they are given below.

$$N_S = 8/\sqrt{26}$$

Herndon[90] has shown recently that

$$\sum_{j \to s} C_{oj} = C_{or} + C_{or'} = ASC(G) \tag{5.27}$$

and

$$C_{or} = |ASC(G_{sr})| \tag{5.28}$$

where $G_{sr} = G_s - r$.

    If we apply eq. (5.27) to anthracene and eq. (5.28) to anthracene α-complex, the following results are obtained:

ASC(anthracene) = 4

ASC(α-complex) = 3

(Note, ASC=K for benzenoid molecules).

Comparison of eqs. (5.25) and (5.28) leads to the formula,

$$[asc(G_s)]^2 = \sum_{j=1}^{N-1} C_{oj}^2 .$$ (5.29)

Substitution of eqs. (5.27) and (5.29) into eq. (5.26) gives,

$$N_s = 2[ASC(G)/asc(G_s)]$$ (5.30)

or by utilizing eq. (5.19)

$$N_s = 2D_s.$$ (5.31)

Therefore, the Dewar reactivity numbers are proportional to the ratio of the algebraic structures count for the ground state and for the reaction intermediate.

Relation (5.31) suggests that the localization energy is a linear function of the logarithm of Dewar number, i.e.,

$$L_r = 2(A + C) + B \ln(N_s/2).$$ (5.32)

In addition, eqs. (5.21) and (5.23), which determine the relative reactivities of two positions, can be rewritten in terms of Dewar number,

$$\Delta L_{rs} = B \ln(N_s/N_r)$$ (5.33)

and

$$k_r/k_s = (N_s/N_r)^\beta$$ (5.34)

Let us evaluate Dewar numbers for biphenylene sigma-complexes using eq. (5.30).

ASC for biphenylene can be obtained using Herndon's method[90] and formula (5.27). The application of Herndon's method consists of the following steps:
(i) Creation of odd-alternant by excising an arbitrary atom from the molecule.

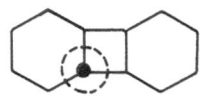

(ii)   Construction of unnormalized NBMO for the obtained odd-alternant fragment.

(iii) Application of eq. (5.27) gives,

ASC(biphenylene) $= C_{or} + C_{or'} + C_{or''} = 3.$

(iv)   In addition, the application of formula

$$\sum_{r \rightarrow s} |C_{or}| = K \qquad (5.35)$$

gives the number of Kekulé structures of biphenylene, K(biphenylene) = 5.

The (asc)-values can be evaluated using eq. (5.29). Let us first construct an NBMO corresponding to graphs G_α and G_β in the way described earlier.

$$[asc(G_\alpha)]^2 = 9$$

$$[asc(G_\beta)]^2 = 12$$

Hence,

$$N(\alpha) = 2(3/\sqrt{9}) = 2$$

$$N(\beta) = 2(3/\sqrt{12}) = 1.73.$$

Since $N(\alpha) > N(\beta)$, the reactivity order is $\alpha < \beta$, which agrees with earlier results.

5.5   NONBONDING MOLECULAR ORBITALS

The appearance of NBMOs in the set of MOs of a conjugated molecule is strongly correlated with its physical and chemical properties. The existence of NBMOs in conjugated molecules led to the prediction[114] that such molecules should have the open-
shell ground states and be extremely reactive. Although in physical reality the situation is much more complicated (for example, because of the Jahn-Teller effect in
the case of triplet ground states), it is experimental fact that structures possessing NBMOs are very rarely found in the chemistry of conjugated hydrocarbons[17] Their
eventual preparation requires drastic experimental conditions[46] Therefore, it is of
importance to establish the presence of NBMOs because their appearance in the set of
molecular orbitals belonging to a particular conjugated structure should be understood as a warning for difficulties which may occur in the preparation of such a
compound.

In the chemical (and mathematical) literature there are different methods available for determining the presence of NBMOs. For example, one approach is to carry
out direct Hückel molecular orbital calculation for a given molecule. However, this
approach does not demonstrate the connections which may exist between NBMOs and topological (structural)* properties of a molecule. Here we wish to describe the topological method for enumerating NBMOs which is a very simple one to use.

The number of NBMOs is identical to the number of zero elements in the graph
spectrum. Since,

$$\det \underline{A} = \prod_{j=1}^{N} x_j \qquad (5.36)$$

the determinant of the adjacency matrix will be zero if and only if there exists at
least one zero in the spectrum of the graph. We have found a simple criterion[29] to
establish whether det $\underline{A}$ is zero or not:

$$\det \underline{A} = (-1)^{N/2} (K^+ - K^-)^2 + (-) ^N \sum_{s \in S_n^*} (-1)^{c(s)} 2^{r(s)}, \qquad (5.37)$$

where S_n^* is the set of Sachs graphs which are simultaneously <u>spanning subgraphs</u> of
a graph and which contain at least one odd cycle. Other symbols appearing in eq.
(5.37) have their previous meaning. A spanning subgraph is the subgraph of a graph
G which contains all vertices of G. For example, a spanning Sachs graph of G is
G_3 in <u>Fig. 5.4.</u>

---

*Molecular topology in a subtle way determines the structure of a conjugated molecule[149]

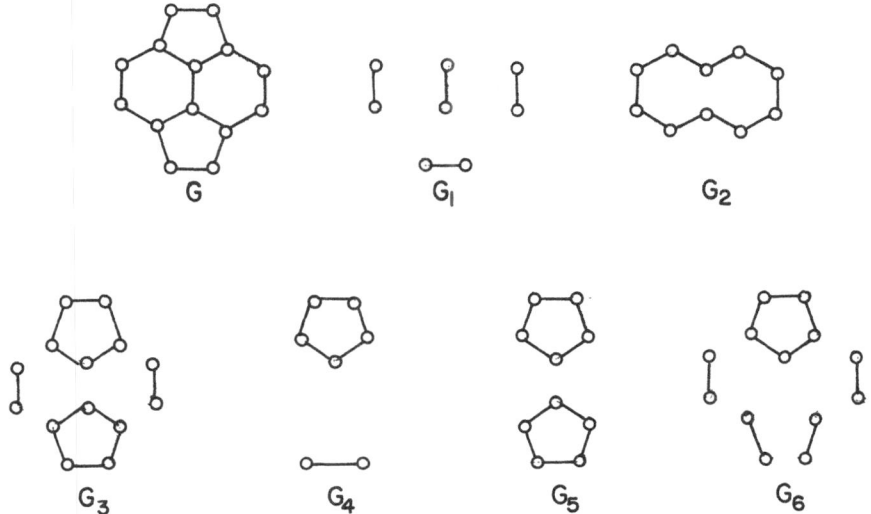

Fig. 5.4. Some subgraphs of a pyracyclene graph.

Formula (5.37) for alternant hydrocarbons (because $S_n^* = \phi$) reduces to

$$\det \underline{A} = (-1)^{N/2} (K^+ - K^-)^2. \qquad (5.38)$$

This is a well-known result of Dewar and Longuet-Higgins[37] which demonstrates the congruency between Hückel theory and resonance theory.

The use of formula (5.37) will be illustrated on pyracyclene (N=14). It consists of the following steps:

(1) We first write down the Kekulé structures and the corresponding Kekulé graphs of pyracyclene.

(a) Kekulé structures

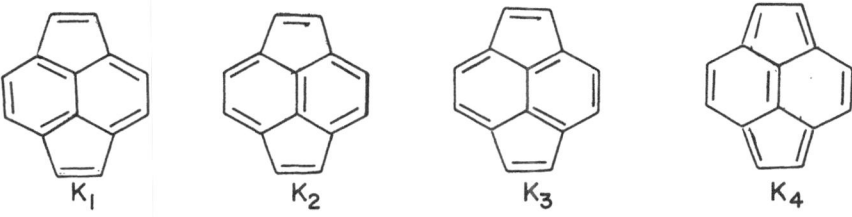

(b) Kekulé graphs (are Sachs graphs with N vertices containing as components only complete graphs of degree one).

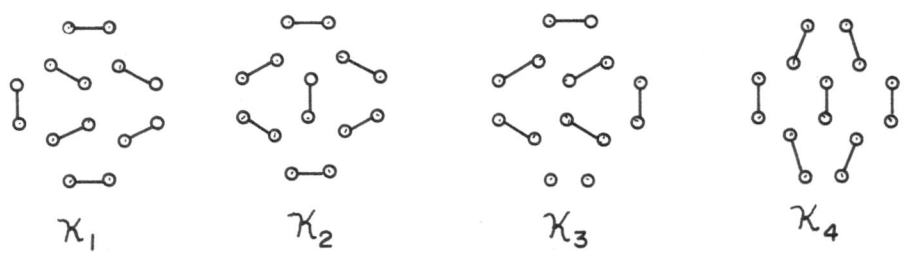

$\mathcal{K}_1$ $\mathcal{K}_2$ $\mathcal{K}_3$ $\mathcal{K}_4$

(ii) Now, we determine parities of Kekulé structures using the superposition method[27,29] K_1 structure is arbitrarily selected to be of even (+1) parity. Then,

$\mathcal{K}_1 + \mathcal{K}_2 =$ $K_2 = +1$

$\mathcal{K}_1 + \mathcal{K}_3 =$ $K_3 = +1$

$\mathcal{K}_1 + \mathcal{K}_4 =$ $K_4 = -1$

Therefore, $K^+ = 3$ and $K^- = 1$.

(iii) Pyracyclene has only one spanning Sachs graph which contains the odd-membered rings. This is G_3 in Fig. 5.4.

$c(s) = 4$
$r(s) = 2$

S_{14}^*

(iv) The evaluation of det $\underline{A}$ by putting (i)-(iii) into expression (5.37):

$$\det \underline{A} = (-1)^{14/2} (3-1)^2 + (-1)^{14} (-)^4 2^2 = 0.$$

Therefore, pyracyclene must have at least one NBMO. The chemistry of pyracyclene appears to be in accord with this graphical prediction; pyracyclene is a very unstable molecule indeed and it could not be isolated from the solution.[113a]

Therefore, the problem whether a conjugated molecule has or has not NBMO is solved. Another problem, however, is how to enumerate NBMOs graphically. The actual number (N_0) of zeros (i.e., NBMOs) in the graph spectrum may be obtained following the Longuet-Higgins-Živković[114,165] algorithm for enumeration of N_0. Let us first consider the eigenvalue equation.

$$\underline{C_i} \underline{A} = x_i \underline{C}. \qquad (5.39)$$

In the case of $N_0 \neq 0$, $x_i = 0$, and $\underline{C_i} = \underline{C_o} \equiv$ NBMO,

$$\underline{C_o} \underline{A} = 0 \qquad (5.40)$$

or in scalar form,

$$\sum_{j \to k} C_{oj}; \quad k = 1, 2, \ldots, N. \qquad (5.41)$$

The summation is over all vertices j joined to the vertex k. This equation has been first used by Longuet-Higgins.[114] It should be also noted[165] that the number of independent parameters in an unnormalized NBMO is equal to N_0. Thus, the enumeration N_0 is reduced to the determination of the number of independent parameters which fulfill requirement (5.41). Application of this method is also illustrated for pyracyclene

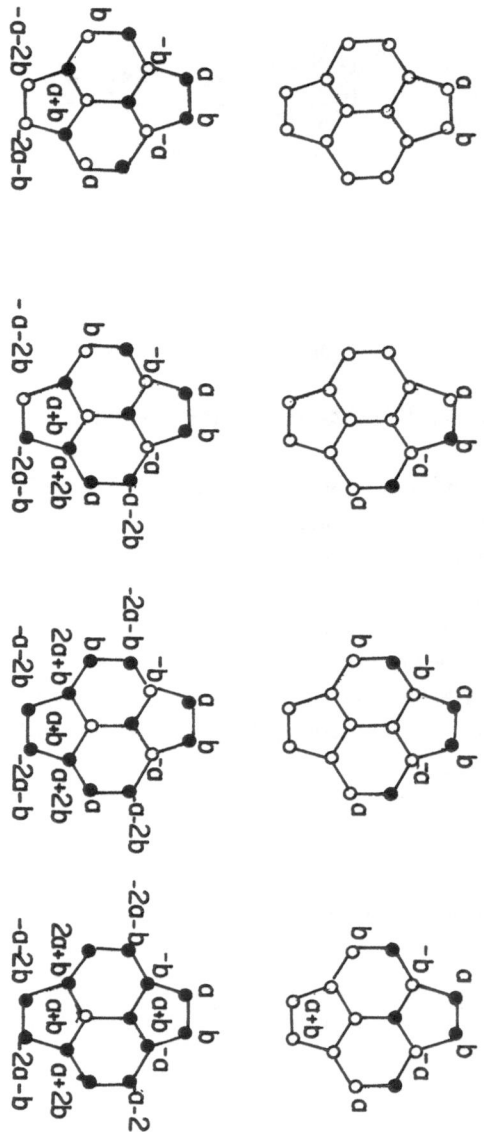

Fig. 5.5. Graphical enumeration of NBMO for pyracyclene.

(see Fig. 5.5.).

Each vertex for which eq. (5.41) is fulfilled is denoted by •. This procedure of fulfilling eq. (5.41) for each vertex is carried in steps. In order that eq. (5.41) is fulfilled for the last (unmarked) vertex of pyracyclene the following relation must hold: a+b = 0. Thus, only one parameter (b=-a) is independent, and consequently N_O=1. The NBMO of pyracyclene is, therefore, given as follows:

$$a = 0.29$$

The whole procedure may, in fact, be simplified by applying such graph trans-formations under which the value of N_O does not change, but which make the graph-theoretical procedure outlined here much easier to apply. The graph transformations are detailed in several works:[26,29,69]

6.   CONCLUSIONS

The topological approach to chemistry in general, and to conjugated structures in particular, has flourished in the last decade for several reasons. We shall list some of them below:

(i) Quantum chemical methods suffer from a common disadvantage because their results lack the generality and simplicity which would enable chemists to apply them in practice to a variety of chemical phenomena.

(ii) Some results obtained from the topological considerations may not be always derived from purely numerical calculations, because the sheer wealth of numerical data which has to be examined may often obscure a simple concept for the understanding or the rationalization of a particular chemical result.

(iii) In addition, a number of results may be obtained by just a pencil and a paper, without any use of computers. This is of importance for experimental chemists who are sometimes hampered in applying current computational theories because of the necessity of carrying out on computers using the specialized programs (not to mention the cost of some of these computations).

(iv) Finally, in developing the topological (mathematical) theory of chemistry our intention is not to compete with quantum-mechanical calculations, but rather to develop a symbolism that permits chemists to think in terms of graphical structures (since this is a natural way of chemical thinking about molecules and reactions); i.e., to learn as much as possible about the eventual chemical behaviour of the molecule by examining the properties of its molecular graph. Furthermore, we also suggest this procedure to those chemists who intend to use quantum-mechanical methods as the preliminary step which could be helpful in guiding and directing their calculations.

Acknowledgements. This work was supported in part by National Science Foundation Grant (GP-42908).

7.   <u>LITERATURE</u>

1.   Badger, G. M.: <u>Aromaticity.</u> Cambridge: University Press. 1962, p. 112
2.   Baird, N.: J. Chem. Educ. <u>48</u>, 509 (1971)
3.   Balaban, A. T.: Rev. Roumaine Chim. <u>15</u>, 1243 (1970)
4.   Balaban, A. T., and F. Harary: J. Chem. Educ. <u>11</u>, 2548 (1971)
5.   Battiste, M.: J. Am. Chem. Soc. <u>86</u>, 942 (1964)
6.   Berge, C.: <u>The Theory of Graphs and Its Applications.</u> London: Methuen 1962
7.   Biggs, N.: <u>Algebraic Graph Theory.</u> Cambridge: University Press 1974
8.   Bloch, E., R. A. Marty and P. de Mayo: J. Am. Chem. Soc. <u>93</u>, 3072 (1971)
9.   Bonchev, D, and N. Trinajstić: J. Chem. Phys., in press
10. Breslow, R. and E. Mohacsi: J. Am. Chem. Soc. <u>85</u>, 431 (1963)
11. Brown, R. D., F. R. Burden, A. J. Jones, and J. E. Kent: J.C.S. Chem. Comm.
     <u>1967,</u> 808
12. Cava, M. P., and M. J. Mitchell: <u>Cyclobutadienes and Related Compounds.</u>
     New York: Academic Press 1967
13. Cava, M. P., M. A. Sprecher, and W. R. Hall: J. Am. Chem. Soc. <u>96</u>, 1817 (1974)
14. Calder, J. G., P. J. Garratt, H. C. Longuet-Higgins, F. Sondheimer, and
     R. Wolovsky: J. Chem. Soc. C <u>1967,</u> 1041
15. Cayley, A.: Phil. Mag. <u>13</u>, 19 (1857)
16. Chapman, O. L., C. C. Chang, and N. R. Rosenquist: J. Am. Chem. Soc. <u>98</u>, 261
     (1976)
17. Clar, E., W. Kemp, and D. C. Stewart: Tetrahedron <u>3</u>, 36 (1958)
18. Collatz, L., and U. Sinogowitz: Abh. Math. Sem. Univ. Hamburg <u>21,</u> 63 (1967)
19. Contrell, T. S.: Tetrahedron Lett. <u>1973</u>, 1853
20. Coulson, C. A.: Proc. Cambridge Phil. Soc. <u>36</u>, 201 (1940)
21. Coulson, C. A.: Proc. Cambridge Phil. Soc. <u>46</u>, 202 (1950)
22. Coulson, C. A.: J. Chem. Soc. <u>1954</u>, 3111
23. Coulson, C. A., and H. C. Longuet-Higgins: Proc. Roy. Soc. (London) <u>A 191,</u>
     39 (1947)
24. Coulson, C. A. and G. S. Rushbrooke: Proc. Cambridge Phil. Soc. <u>36</u>, 193 (1940)
25. Coulson, C. A. and A. Streitwieser, Jr.: <u>Dictionary of Π-Electron Calculations.</u>
     San Francisco: Freeman 1965
25a. Cvetković, D.: Publ. Fac. Electrotechnique Univ. Belgrade <u>354-356</u>, 1 (1971)
26. Cvetković, D., I. Gutman, and N. Trinajstić: Croat. Chem. Acta <u>44</u>, 365
     (1972)
27. Cvetković, D., I. Gutman, and N. Trinajstić: J. Chem. Phys. <u>61</u>, 2700 (1974)
28. Cvetković, D., I. Gutman, and N. Trinajstić: Chem. Phys. Lett. <u>29</u>, 65 (1974)
29. Cvetković, D., I. Gutman, and N. Trinajstić: J. Mol. Struct. <u>28</u>, 289 (1975)
30. Dauben, H. J. and D. J. Bertelli: J. Am. Chem. Soc. <u>83</u>, 4659 (1961)
31. Dewar, M. J. S.: Chem. Eng. News 43, 86 (1965); Tetrahedron Suppl. <u>8</u>, 75 (1966)

32. Dewar, M. J. S.: The Molecular Orbital Theory of Organic Chemistry. New York: McGraw-Hill 1969

33. Dewar, M. J. S.: Angew. Chem. Internat. Edit. 10, 761 (1971)

33a. Dewar, M. J. S., and R. C. Dougherty: The PMO Theory of Organic Chemistry. New York: Plenum Press 1975

34. Dewar, M. J. S. and C. deLlano: J. Am. Chem. Soc. 91, 789 (1969)

35. Dewar, M. J. S., R. C. Haddon, and P. J. Student: J.C.S. Chem. Comm. 1974, 569

36. Dewar, M. J. S., A. J. Harget, and N. Trinajstić: J. Am. Chem. Soc. 91, 6321 (1969)

37. Dewar, M. J. S., and H. C. Longuet-Higgins: Proc. Roy. Soc. (London) A 214, 482 (1952)

38. Dewar, M. J. S., and R. J. Sampson: J. Chem. Soc. 1956, 2789

39. Dewar, M. J. S., and N. Trinajstić: Coll. Czech. Chem. Comm. 35, 3136, 3484 (1970)

40. Dziewoński, K. and G. Rapalski: Ber. 45, 2941 (1912)

41. Eckert, J. M.: J. Chem. Educ. 7, 458 (1973)

42. England, W., and K. Ruedenberg: J. Am. Chem. Soc. 95, 8769 (1973)

43. Essam, J. W., and M. E. Fischer: Rev. Mod. Phys. 42, 272 (1970)

44. Euler, L.: Comment. Academiae Sci. Petropolitanae 8, 128 (1736)

45. Farquhar, D., T. T. Gough, and D. Leaver: J.C.S. Perkin I 1976, 341

46. e.g., Flynn, C. R., and J. Michl: J. Am. Chem. Soc. 95, 5802 (1973)

47. Frisch, H. L., and E. J. Wasserman: J. Am. Chem. Soc. 83, 3789 (1961)

48. Gantmacher, F. R.: Theory of Matrices. Moscow: Nauka 1967

49. Garratt, P. J., and M. V. Sargent: Advances in Organic Chemistry. Edited by E. C. Taylor and H. Wynberg. New York: Wiley 1969, p. 1

50. Garratt, P. J.: Aromaticity. London: McGraw-Hill, 1971

51. Graovac, A., I. Gutman, and N. Trinajstić: Chem. Phys. Lett. 35, 555 (1975)

51a. Graovac, A., I. Gutman, and N. Trinajstić: Chem. Phys. Lett. 37, 471 (1976)

52. Graovac, A., I. Gutman, N. Trinajstić, and T. Živković: Theoret. Chim. Acta 26, 67 (1972)

53. Graovac, A., and N. Trinajstić: Croat. Chem. Acta 47, 95 (1975)

54. Graovac, A., and N. Trinajstić: J. Mol. Struct. 30, 416 (1976)

54a. Graovac, A., O.E. Polansky, N. Tyutyulkov, and N. Trinajstić: Z. Naturforsch. 30a, 1696 (1975)

55. Gronowitz, S., U. Rudén, and B. Gestblom: Arkiv Kemi. 20, 297 (1963)

56. von Gustorf, E.K., M.C. Henry, and P.V. Kennedy: Angew. Chem. Internat. Edit. 6, 627 (1967)

57. Gutman, I.: Croat. Chem. Acta 46, 209 (1974)

58. Gutman, I.: Chem. Phys. Lett. 24, 283 (1974)

59. Gutman, I.: Theoret. Chim. Acta 35, 355 (1974)

60. Gutman, I.: Croat. Chem. Acta 48, 97 (1976)

61. Gutman, I., J. V. Knop, and N. Trinajstić: Z. Naturforsch. 29b, 80 (1974)

62. Gutman, I., M. Milun, and N. Trinajstić: Croat. Chem. Acta 44, 207 (1972)

63. Gutman, I., M. Milun, and N. Trinajstić: J. Chem. Phys. 59, 2772 (1973)

64. Gutman, I., M. Milun, and N. Trinajstić: Croat. Chem. Acta 48, 87 (1976)

65. Gutman, I., M. Milun, and N. Trinajstić: J. Am. Chem. Soc. 99, 1692 (1977)

66. Gutman, I., B. Ruščić, N. Trinajstić, and C. F. Wilcox, Jr.: J. Chem. Phys. 62, 3399 (1975)

67. Gutman, I., and N. Trinajstić: Chem. Phys. Lett. 17, 535 (1972)

68. Gutman, I., and N. Trinajstić: Topics Curr. Chem. 42, 49 (1973)

69. Gutman, I., and N. Trinajstić: Croat. Chem. Acta 45, 423 (1973); ibid. 45, 539 (1973)

70. Gutman, I., and N. Trinajstić: Chem. Phys. Lett. 20, 257 (1973)

71. Gutman, I., and N. Trinajstić: Z. Naturforsch. 29a, 1238 (1974)

72. Gutman, I., and N. Trinajstić: Croat. Chem. Acta 47, 95 (1975)

73. Gutman, I., and N. Trinajstić: Croat. Chem. Acta 47, 507 (1975)

74. Gutman, I., and N. Trinajstić: Can.J. Chem. 54, 1789 (1976)

74a. Gutman, I., and N. Trinajstić: J. Chem. Phys. 64, 4921 (1976)

75. Gutman, I., N. Trinajstić, and C. F. Wilcox, Jr.: Tetrahedron 31, 143 (1975)

76. Gutman, I., N. Trinajstić, and T. Živković: Croat. Chem. Acta 44, 501 (1972)

77. Gutman, I., N. Trinajstić, and T. Živković: Chem. Phys. Lett. 14, 342 (1972)

78. Gutman, I., N. Trinajstić, and T. Živković: Tetrahedron 29, 3449 (1973)

78a. Günthard, H. H., and H. Primas: Helv. Chim Acta 39, 1645 (1956)

79. Haddon, R. C., V. R. Haddon, and L. M. Jackman: Topics Curr. Chem. 16, 103 (1971)

80. Hafner, A.: Angew. Chem. 75, 1041 (1963)

81. Hakala, R. W.: Internat. J. Quantum Chem. 1S, 187 (1967)

82. Hall, G. G.: Proc. Roy. Soc. (London) A 229, 251 (1955)

83. Hall, G. G.: Internat. J. Math. Educ. Sci. Technol. 4, 233 (1973)

84. Ham, N. S.: J. Chem. Phys. 29, 1229 (1958)

85. Harary, F.: Graph Theory. Reading (mass.): Addison-Wesley 1971, Second printing

86. Harary, F.: Uspekhi Mat. Nauk 24, 179 (1969)

87. Heilbronner, E.: Helv. Chim. Acta 36, 170 (1953)

88. Heilbronner, E.: Helv.Chim. Acta 37, 921 (1954)

89. Heilbronner, E.: Tetrahedron Lett. 1923 (1964)

90. Herndon, W.C.: Tetrahedron 29, 3 (1973)

90a. Herndon, W.C.: Tetrahedron Lett. 671 (1974)

91. Herndon, W.C.: J. Chem. Educ. 51, 10 (1974)

91a. Herndon, W.C., and M.L. Ellzey, Jr.: Tetrahedron 31, 99 (1975)

92. Hess, B. A., Jr. and L. J. Schaad: J. Amer. Chem. Soc. 93, 305, 2413 (1971)

93. Hess, B. A., Jr. and L. J. Schaad: J. Org. Chem. 36, 3418 (1971)

94. Hess, B. A., Jr., and L. J. Schaad: J. Am. Chem. Soc. 95, 3907 (1973)

95. Hess, B. A., Jr., L. J. Schaad, and C. W. Holyoke, Jr.: Tetrahedron 28, 3657 (1972); ibid. 31, 295 (1975)

96. Hoffmann, J. M., Jr., and R. H. Schlessinger: J. Am. Chem. Soc. 92, 5203 (1970)

97. Hosoya, H.: Bull. Chem. Soc. Japan 44, 2332 (1971)

98. Hosoya, H.: Theoret. Chim. Acta 25, 215 (1972)

99. Hosoya, H., K. Hosoi, and I. Gutman: Theoret. Chim. Acta 38, 37 (1975)

100. Hückel, E.: A. Physik 70, 204 (1931); ibid. 72, 310 (1931)

101. Hückel, E.: Z. Physik 76, 628 (1932)

102. Huntsman, W. D. and H. J. Wristers. J. Am. Chem. Soc. 89, 342 (1967)

103. Isaacs, N. S.: Reactive Intermediates in Organic Chemistry. New York: Wiley 1974

103a. Jotham, R. W.: Chem. Soc. Revs. 2, 457 (1973)

104. Katz, T. J.: J. Am. Chem. Soc. 82, 3784, 3785 (1960)

105. Keller-Schierlein, W., and E. Heilbronner: In: Non-Benzenoid Hydrocarbons. Edited by D. Ginsburg. New York: Interscience 1959, p. 277

106. Kelly, P. J.: Pacific J. Math. 7, 961 (1957)

107. Kende, A. S., P. T. Izzo, and P. T. MacGregor: J. Am. Chem. Soc. 88, 3359 (1966)

107a. Kershner, L. D., J. M. Goudis, and H. H. Freedman: J. Am. Chem. Soc. 94, 985 (1972)

108. König, D.: Theorie der endlichen und unendlichen Graphen. Leipzig 1936

109. Kruszewski, J., and T. M. Krygowski: Can. J. Chem. 53, 945 (1975)

110. Kuratowski, K.: Fund. Math. 15, 271 (1930)

111. Kurosh, A. G.: Kurs vishei algebri. Nauka 1965

112. Krantz, A., C. Y. Lin, and M. D. Newton: J. Am. Chem. Soc. 95, 2744 (1973)

113. Lin, C. Y., and A. Krantz: J.C.S. Chem. Comm. 1972, 1111

113a. Lloyd, D.: Carbocyclic Non-Benzenoid Aromatic Compounds. Amsterdam: Elsevier 1966

114. Longuet-Higgins, H. C.: J. Chem. Phys. 18, 265 (1950)

115. Lovácz, L., and J. Pelikán: Period. Math. Hung. 3, 215 (1972)

116. Maier, G.: Angew. Chem. Internat. Edit. 6, 402 (1967)

117. Manvel, B.: In: Proof Techniques in Graph Theory. Edited by F. Harary. New York: Academic Press 1969

118. Marcus, R. A.: J. Chem. Phys. 43, 2643 (1965)

119. McClelland, B. J.: J. Chem. Phys. 54, 640 (1971)

120. Mowshowitz, A.: J. Combinatorial Theory 12 (B), 177 (1972)

121. Murrel, J. N., S. F. A. Kettle, and J. M. Tedder: Valence Theory. London: Wiley 1965

121a. Nosal, E.: Eigenvalues of Graphs. Ph.D. Thesis. The University of Calgary: Calgary 1970

122. Oberschelp, W.: Math. Ann. 174, 53 (1967)

123. Olah, G., J. S. Staral, and L. A. Paquette: J. Am. Chem. Soc. 98, 1267 (1976)

124. Platt, J. R.: J. Chem. Phys. 15, 419 (1947); ibid. 56, 328 (1962)

125. Polya, G.: Acta Math. 68, 145 (1937)

126. Polansky, O. E.: Mathematical Chemistry 1, 183 (1975)

127. Prelog, V.: Nobel Lecture, December 12th, 1975

128. Randić, M.: J. Chem. Phys. 60, 3920 (1974)

129. Randić, M.: J. Am. Chem. Soc. 97, 6609 (1975)

130. Randić, M., N. Trinajstić, and T. Živković: J.C.S. Faraday Tran II 1976, 244

131. Riordan, J.: An Introduction to Combinatorial Analysis. New York: Wiley 1958

132. Rouvray, D. H.: Compt. Rend. 275C, 657 (1972)

133. Rouvray, D. H.: Amer. Sci. 61, 729 (1973)

134. Rouvray, D. H.: Chem. Soc. Revs. 3, 355 (1974)

135. Ruedenberg, K.: J. Chem. Phys. 22, 1878 (1954)

136. Ruedenberg, K.: J. Chem. Phys. 29, 1232 (1958)

137. Ruedenberg, K.: J. Chem. Phys. 34, 1884 (1961)

138. Sachs, H.: Publ. Math. (Debrecen) 11, 119 (1964)

139. Schaad, L. J., and B. A. Hess, Jr.: J. Am. Chem. Soc. 94, 3068 (1972)

139a. Schmidtke, H. H.: J. Chem. Phys. 45, 3920 (1966)

140. Sondheimer, F.: Pure Appl. Chem. 7, 303 (1969)

141. Sondheimer, F.: Accounts Chem. Res. 5, 81 (1972)

142. Sondheimer, F., R. Wolovsky, and Y. Amiel: J. Am. Chem. Soc. 84, 274 (1962)

143. Spialter, L.: J. Chem. Docum. 4, 261, 269 (1964)

144. Stang, P. J., and M. G. Magnum: J. Am. Chem. Soc. 97, 3854 (1975)

145. Stepanov, N. F. and V. M. Tatevskii: Zhur. Strukt. Khim. 2, 241, 452 (1961)

146. Streitwieser, A., Jr.: Molecular Orbital Theory for Organic Chemists. New York: Wiley 1961

147. Sussenguth, E. H., Jr.: J. Chem. Docum. 5, 36 (1965)

148. Sylvester, J. J.: Amer. J. Math. 1, 87 (1878)

149. Trinajstić, N.: In: Modern Theoretical Chemistry - Semiempirical Methods of Electronic Structure Calculations. Part A: Techniques. Ed. G. A. Segal. Volume 7. New York: Plenum 1977, p. 1

150. Turner, J.: SIAM J. Appl. Math. 16, 520 (1968)

151. Ulam, S. M.: A Collection of Mathematical Problems. New York: Wiley 1960

152. Veblen, O.: Analysis Situs. Am. Math. Soc. Colloq. Publ. Cambridge. Volume 5

153. Wheland, G. W.: J. Am. Chem. Soc. 64, 900 (1942)

154. Wheland, G. W.: The Theory of Resonance and Its Application to Organic Chemistry. New York: Wiley 1953

155. Wiener, H.: J. Am. Chem. Soc. 69, 17, 2636 (1947)

156. Wilcox, C. F., Jr.: Tetrahedron Lett. 1968, 795

157. Wilcox, C. F., Jr.: J. Am. Chem. Soc. 91, 2732 (1969)

158. Wilcox, C. F., Jr.: Croat. Chem. Acta <u>47</u>, 98 (1975)

159. Wilcox, C. F., Jr., I. Gutman, and N. Trinajstić: Tetrahedron <u>31</u> 147 (1975)

160. Windgassen, R., Jr., W. H. Saunders, Jr. and V. Boekelheide: J. Am. Chem. Soc. <u>81</u>, 1459 (1959)

161. Winter, R. E. K.: Tetrahedron Lett. <u>1965</u>, 1207

162. Wynberg, H. and D. J. Zwanenburg: Tetrahedron Lett. <u>1967</u>, 761

163. Zimmermann, H. E.: J. Am. Chem. Soc. <u>88</u>, 1564, 1566 (1966)

164. Zykov, A. A.: <u>Teoriya Konechnykh grafov.</u> Novosibirsk 1969

165. Živković, T.: Croat. Chem. Acta <u>44</u>, 351 (1972)

166. Živković, T., N. Trinajstić and M. Randić: Mol. Phys. <u>30</u>, 517 (1975)

SUBJECT INDEX